Advanced Topics in Science and Technology in China

Volume 60

Zhejiang University is one of the leading universities in China. In Advanced Topics in Science and Technology in China, Zhejiang University Press and Springer jointly publish monographs by Chinese scholars and professors, as well as invited authors and editors from abroad who are outstanding experts and scholars in their fields. This series will be of interest to researchers, lecturers, and graduate students alike.

Advanced Topics in Science and Technology in China aims to present the latest and most cutting-edge theories, techniques, and methodologies in various research areas in China. It covers all disciplines in the fields of natural science and technology, including but not limited to, computer science, materials science, the life sciences, engineering, environmental sciences, mathematics, and physics. This book series is indexed by the SCOPUS database.

If you are interested in publishing your book in the series, please contact Dr. Mengchu Huang(Email: mengchu.huang@springer.com).

More information about this series at http://www.springer.com/series/7887

Yong Ding · Guangming Sun

Stereoscopic Image Quality Assessment

Yong Ding
College of Information Science
and Electronic Engineering
Zhejiang University
Hangzhou, Zhejiang, China

Guangming Sun
College of Information Science
and Electronic Engineering
Zhejiang University
Hangzhou, Zhejiang, China

ISSN 1995-6819　　　　　　ISSN 1995-6827　(electronic)
Advanced Topics in Science and Technology in China
ISBN 978-981-15-7763-5　　　ISBN 978-981-15-7764-2　(eBook)
https://doi.org/10.1007/978-981-15-7764-2

Jointly published with Zhejiang University Press, China
The print edition is not for sale in China (Mainland). Customers from China (Mainland) please order the print book from: Zhejiang University Press.

© Zhejiang University Press 2020
This work is subject to copyright. All rights are reserved by the Publishers, whether the whole or part of the material is concerned, specifically the rights of translation, reprinting, reuse of illustrations, recitation, broadcasting, reproduction on microfilms or in any other physical way, and transmission or information storage and retrieval, electronic adaptation, computer software, or by similar or dissimilar methodology now known or hereafter developed.
The use of general descriptive names, registered names, trademarks, service marks, etc. in this publication does not imply, even in the absence of a specific statement, that such names are exempt from the relevant protective laws and regulations and therefore free for general use.
The publishers, the authors, and the editors are safe to assume that the advice and information in this book are believed to be true and accurate at the date of publication. Neither the publishers nor the authors or the editors give a warranty, express or implied, with respect to the material contained herein or for any errors or omissions that may have been made. The publishers remain neutral with regard to jurisdictional claims in published maps and institutional affiliations.

This Springer imprint is published by the registered company Springer Nature Singapore Pte Ltd.
The registered company address is: 152 Beach Road, #21-01/04 Gateway East, Singapore 189721, Singapore

Preface

With the rapid development of digital image and video acquisition, transmission and display techniques in the last few decades, the demands of high-quality images and videos are growing amazingly fast in both people's everyday lives and specific application scenarios such as the fields of academy and engineering. Over the past years, the wave of stereoscopic display technology has an exponential increase, which can be mainly witnessed by the huge number of high-quality three-dimensional (3D) films and 3D TV in large-scale application. Meanwhile, mobile phones are also expected to be the largest 3D display application in the near future. There is no denial that the rapid development of stereoscopic technology has significantly enriched the way people perceive the world psychologically. Nevertheless, during capturing, coding, transmitting, processing and displaying of stereoscopic images, the distortion and interference introduced from outside are inevitable and unneglectable, which lead to the decrease of image quality and great visual discomfort. Thus, it is highly necessary to design effective methods to evaluate the perceptual quality of images, which is of vital importance for the performance optimization of image processing systems.

However, due to the special characteristics of stereoscopic images that are distinguished from two-dimensional (2D) images, for example, complex and non-intuitive interactions between multiple 3D visual cues, including depth perception, visual comfort and binocular characteristics such as binocular fusion and rivalry, automatically assessing the quality of stereoscopic images is still a challenging issue.

In recent years, although a large number of experimental studies on stereoscopic image quality assessment have been performed and various factors that affect stereoscopic perception have been investigated, it is still a puzzle in fully understanding the neural mechanism of visual cortex about how the human brain perceives and deals with stereoscopic natural image. As a result, studying upon stereoscopic image quality assessment is attracting more and more attention and a significant progress has been witnessed.

This book attempts to discuss the related topics about stereoscopic image quality assessment thoroughly and systematically. Firstly, the difference between 2D and stereoscopic image quality assessment is given. Secondly, the research of stereoscopic image quality assessment is discussed and analysed detaily, including a straightforward way based on existing 2D methods, a perceptual way based on human visual system properties and a new trend making use of deep learning models. Finally, some emerging challenges are described, and meanwhile a few new directions and trends are explored that are worth further investigations and research.

The authors would give particular thanks to Xiaoshu Xu who has made a significant contribution to the publication of this book. Moreover, the authors would express deep appreciation to the other students of Prof. Ding who have contributed to the research works presented in this book. For example, Ruizhe Deng, Yang Zhao, Xiaogang Xu, Zijin Gu, et al. have made their great efforts in researching on stereoscopic image quality assessment for the last several years.

Besides, the authors have received generous assistance and support from many of our colleagues including valuable information and materials used in this book, discussions, feedback, comments on and proofreading of various parts of the book, recommendations and suggestions that shaped the book as it is.

Due to our limited knowledge and energy, there inevitably exist some ambiguous interpretations and even mistakes in this book, which we welcome the readers and colleagues to point out.

Hangzhou, China Yong Ding
June 2020

Contents

1	**Introduction**	1
	References	4
2	**Brief Introduction on 2D Image Quality Assessment**	7
	2.1 Introduction	7
	2.1.1 Public Image Quality Databases	9
	2.1.2 IQA Algorithm Performance Metrics	10
	2.1.3 Typical Objective IQA Algorithms	13
	2.2 Summary	27
	References	28
3	**Difference Between 2D and Stereoscopic Image Quality Assessment**	31
	3.1 Introduction	31
	3.2 Binocular Vision	34
	3.2.1 Binocular Disparity	34
	3.2.2 Binocular Fusion	35
	3.2.3 Binocular Rivalry	36
	3.2.4 Ocular Dominance	37
	3.2.5 Visual Discomfort	38
	3.3 Subjective Stereoscopic Image Quality Assessment	39
	3.3.1 Principle	39
	3.3.2 Databases	40
	3.4 General Frameworks of SIQA Models	43
	3.5 Summary	45
	References	46
4	**SIQA Based on 2D IQA Weighting Strategy**	49
	4.1 Introduction	49
	4.2 SIQA Algorithms Based on 2D IQA Methods	50
	4.3 SIQA Algorithms Employ the Disparity Information	51
	4.4 The State-of-art SIQA Algorithms Based 2D IQA Weighting	59

	4.5	Summary	65
	References		65
5	**Stereoscopic Image Quality Assessment Based on Binocular Combination**		**69**
	5.1	Introduction	69
	5.2	How to Form the Cyclopean Image	71
	5.3	Region Classification Strategy	78
	5.4	Visual Fatigue and Visual Discomfort	83
		5.4.1 Visual Fatigue Prediction for Stereoscopic Image	83
		5.4.2 Visual Discomfort Prediction for Stereoscopic Image	86
	5.5	Summary	92
	References		93
6	**Stereoscopic Image Quality Assessment Based on Human Visual System Properties**		**97**
	6.1	Introduction	97
	6.2	Human Visual System	99
	6.3	SIQA Based on Hierarchical Structure	101
	6.4	SIQA Based on Visual Saliency	105
		6.4.1 Visual Saliency Models	105
		6.4.2 Application of Visual Saliency in 3D IQA	110
	6.5	SIQA Based on Just Noticeable Difference	118
		6.5.1 Just Noticeable Difference	118
		6.5.2 Application of JND in 3D IQA	123
	6.6	Summary	127
	References		129
7	**Stereoscopic Image Quality Assessment Based on Deep Convolutional Neural Models**		**135**
	7.1	Introduction	135
	7.2	Stereoscopic Image Quality Assessment Based on Machine Learning	137
	7.3	Stereoscopic Image Quality Assessment Based on Transfer Learning	139
		7.3.1 Theoretical Basis for Transfer Learning	139
		7.3.2 From Image Classification to Quality Regression Task	140
		7.3.3 From Image Classification to Quality Classification Task	141
	7.4	Stereoscopic Image Quality Assessment Based on Patch-wise Models	143
		7.4.1 Theoretical Basis for Patch-wise Strategy	143
		7.4.2 Patch-wise Strategy with Global Subjective Score	145
		7.4.3 Patch-wise Strategy with Generated Quality Map	146
		7.4.4 Saliency-guided Local Feature Selection	147
		7.4.5 Dual-stream Interactive Networks	149

	7.5	New Tendency for Exploiting CNN-Based SIQA Tasks	151
	7.6	Other Necessary Knowledge in CNN-Based SIQA Tasks	152
		7.6.1 Image Preprocessing	152
		7.6.2 Activation Function	153
		7.6.3 Loss Function	154
		7.6.4 Regularization	155
		7.6.5 Optimization	157
		7.6.6 Summary	158
	7.7	Summary and Future Work	158
	References		160
8	**Challenging Issues and Future Work**		165

Chapter 1
Introduction

Abstract Nowadays, objective image quality assessment (IQA) plays an important role for performance evaluation of image/video processing systems. Over the past few years, a variety of IQA methods have been introduced and they can be divided into three categories: full-reference IQA, reduced-reference IQA and no-reference IQA. All of these methods are clarified in detail in this book. In this chapter, the overall structure of the book is explained briefly and a summary of each of the following chapters is also provided.

Keywords Image quality assessment · Performance evaluation · Stereoscopic image

The idiom saying "A picture us worth a thousand words" has illustrated the importance of visual information perceived by human beings from nature images. With the rapid development of image store and display technologies, the image applications are widely used in human daily life, including entertainment, communications, security, monitoring and medical treatment fields (Wang and Bovik 2009; Karam et al. 2009). Perceptual quality of images plays an essential role in human perceiving visual information by human brain. Unfortunately, various distortions could be introduced during image transmission, compression, encoding and decoding, leading to images suffering from potentially substantial loss (Larson and Chandler 2010). Therefore, it is urgent demand to monitor and evaluate quality of images in real-time, which is called image quality assessment (IQA). In the past few years, there are a number of studies on IQA research community, and several of them have achieved promising results. Recently, three-dimensional (3D) media applications like virtual reality (VR) and 3D television (3D-TV) have been invented to improve the quality of human life, which derives consumers' interests for high-quality image contents of stereo images instead of plane images (Shen et al. 2018). How to explore the perceptual quality of stereo images is becoming the focus of research in recent years (Moorthy et al. 2013).

Since nature images are perceived by human eyes and processed in human brain, the quality score of distorted images can be obtained in subjective test accurately. However, it is time-consuming and laborious, and cannot realized in real time in uncontrolled environments (Wang et al. 2004). It triggers the urgent demand for developing real-time reliable objective IQA methods to explore the quality of images, whose predicted quality scores are expected to be consistent with subjective perceptual scores by human eyes.

Further, objective IQA methods can be classified into three fields, full-reference (FR), reduced-reference (RR) and no-reference (NR), in which the classification criterion is based on the principle whether reference image information is involved in IQA research. When providing original reference images for comparison, viewers or objective algorithms can better explore the perceptual quality of images, which is called FR IQA. In contrast, NR IQA is defined by predicting quality scores of distorted images without any corresponding reference image information. As a trade-off between FR and NR IQA, RR IQA is conducted with the assistance of reference images partly. Due to the unavailability of reference images in practice applications, the research of NR IQA is the most valuable and challenging among the three.

According to the different inputs, IQA also can be divided into two research directions: 2D IQA for plane images and 3D IQA (also named SIQA) for stereopairs. In the past few years, significant progress has been achieved in IQA research of both 2D and 3D images. Since the main target of this book is to evaluate the quality of stereoscopic images, so most of chapters focus on 3D IQA research community. Of course, because 2D IQA is fundamental of exploring SIQA, we give briefly an overview about IQA research for plane images before introducing 3D IQA research, and interested readers can refer to relevant books and papers.

This book provides a comprehensive survey covering major developments in the field of SIQA. Firstly, the development of SIQA research community is introduced systematically. Then more detailed discussions and analysis in particular methods are presented in each chapter, respectively. Finally, some emerging challenges are described when conducting the research on 3D IQA, and meanwhile a few new directions and trends are explored and discussed that are worth further investigations and research in the future.

For each chapter, we assay progress focusing on a particular way to assess the quality of stereopairs. Each chapter begins with a brief introduction related to the chapter title, also includes an overview of recent significant progress in such aspect of SIQA research, hoping to help readers better understand. Besides, numerous references are given at the end of each chapter, which are from recent classical relative works.

Chapter 2 gives a comprehensive overview of IQA for plane images briefly. The research of 2D IQA is fundamental of exploring 3D images. It is necessary to understand relative knowledges about IQA before introducing SIQA research formally. Therefore, this chapter introduces some of the IQA information necessary for the transition to 3D IQA research community, including public subjective 2D IQA databases and well-known IQA methods (Wang et al. 2004; Winkler 2012). Of course, not all aspects of such a large subject about 2D IQA can be covered only in one single chapter. Unfortunately, limited of space, more detailed information in 2D IQA refer to relative books and papers, where we recommend reading the book titled with "Visual quality assessment for natural and medical image" (Ding 2018).

Chapter 3 formally introduces the conception of SIQA, including subjective ratings and public SIQA databases. In addition, the different visual information with the plane image, called binocular vision (Fahle 1987; Howard and Rogers 1995), is also discussed in this chapter briefly, which can give a illustrate understanding for readers.

Chapter 4 provides the general framework of SIQA based on 2D IQA models weighting. Directly extending some well-known 2D IQA models into SIQA tasks is a preliminary exploration for assessing the quality of stereopairs. In later research, as one of most important 3D visual factors, disparity map is also considered to assist quality prediction for stereoscopic images. Perhaps such a framework is not perfect from today's perspective, but it is necessary for us to learn and understand the contents of this chapter as a pioneer of SIQA research.

Chapter 5 illustrates the binocular vision caused by stereo images in detail. When the two views of stereopairs are combined into stereopsis, binocular visual properties (e.g., depth perception, binocular rivalry and visual discomfort) will occur, especially for asymmetrically distorted stereopairs. Deep analysis about binocular vison is given in this chapter briefly, and some relative works on SIQA fields considering binocular properties are introduced and discussed subsequently.

Chapter 6 focus on the importance of human visual system (HVS) on image quality evaluation. Since the HVS is the receiver of visual information for nature images, simulating the properties of HVS is a meaningful and valid for prediction performance improvements (Krüger et al. 2013). There are many visual properties in HVS that have been explored in previous research, from which visual saliency (Zhang et al. 2014) and just noticeable difference (Legras et al. 2004) are selected to discuss in detail in this chapter, respectively.

Chapter 7 gives a new trend for SIQA research community. Recently, deep learning has applied into many image processing tasks and achieved promising results than before. Many researchers begin to attempt to employ convolutional neural networks (CNNs) into SIQA fields, expecting CNN can automatically learning visual representations related with image quality rather than using hand-crafted visual features. However, the biggest obstacle, inadequate training data, need to be addressed before designing more complex deep learning models. There are many strategies applied in CNN-based SIQA models for alleviating the problem, including patch-wise (Zhang et al. 2016; Oh et al. 2017), transfer learning (Ding et al. 2018; Xu et al. 2019) and extending datasets (Liu et al. 2017; Dendi et al. 2019), which derive a series of SIQA models using CNN architectures.

Chapter 8 gives a summary of SIQA research community described in previous chapters, and meanwhile discusses the challenge issues and new trends of stereoscopic image quality assessment in the future.

The book is intended for researchers, engineers as well as graduate students working on related fields including imaging, displaying and image processing, especially for those who are interested in the research of SIQA. It is believed that the review and presentation of the latest advancements, challenges, and new trends in the stereoscopic image quality assessment will be helpful to the researchers and readers of this book.

References

Dendi SVR, Dev C, Kothari N, Channappayya SS (2019) Generating image distortion maps using convolutional autoencoders with application to no reference image quality assessment. IEEE Signal Process Lett 26(1):89–93

Ding Y (2018) Visual quality assessment for natural and medical image. Springer. https://doi.org/10.1007/978-3-662-56497-4

Ding Y, Deng R, Xie X, Xu X, Chen X et al (2018) No-reference stereoscopic image quality assessment using convolutional neural network for adaptive feature extraction. IEEE Access 6:37595–37603

Fahle M (1987) Two eyes, for what? Naturwissenchaften 74(8):383–385

Howard IP, Roger BJ (1995) Binocular vision and stereopsis. Oxford University Press, New York

Karam LJ, Ebrahimi T, Hemami SS, Pappas TN (2009) Introduction to the issue on visual media quality assessment. IEEE J Sel Topics Signal Process 3(2):189–190

Krüger N, Janssen P, Kalkan S, Lappe M, Leonardis A et al (2013) Deep hierarchies in the primate visual cortex: what can we learn for computer vision? IEEE Trans Pattern Anal Mach Intell 35(8):1847–1871

Larson EC, Chandler DM (2010) Most apparent distortion: full-reference image quality assessment and the role of strategy. J Electron Imaging 19(1):1–21

Legras R, Chanteau N, Charman WN (2004) Assessment of just-noticeable differences for refractive errors and spherical aberration using visual simulation. Optom Vis Sci 81(9):718–728

Liu X, van de Weijer J, Bagdanov AD (2017) RankIQA: learning from rankings for no-reference image quality assessment. International Conference on Computer Vision, Venice, Italy, pp 1040–1049

Moorthy AK, Su CC, Mittal A, Bovik AC (2013) Subjective evaluation of stereoscopic image quality. Sig Process Image Commun 28(8):870–883

Oh H, Ahn S, Kim J, Lee S (2017) Blind deep S3D image quality evaluation via local to global feature aggregation. IEEE Trans Image Process 26(10):4923–4935

Shen L, Li K, Feng G, An P, Liu Z (2018) Efficient intra mode selection for depth-map coding utilizing spatiotemporal, inter-view correlations in 3D-HEVC. IEEE Trans Image Process 27(9):4195–4206

Wang Z, Bovik AC (2009) Mean squared error: love it or leave it?—A new look at signal fidelity measures. IEEE Signal Process Mag 1:98–117

Wang Z, Bovik AC, Sheikh HR, Simoncelli EP (2004) Image quality assessment: from error visibility to structural similarity. IEEE Trans Image Process 13(4):600–612

Winkler S (2012) Analysis of public image and video databases for quality assessment. IEEE J Sel Topics Signal Process 6(6):616–625

Xu X, Shi B, Gu Z, Deng R, Chen X et al (2019) 3D no-reference image quality assessment via transfer learning and saliency-guided feature consolidation. IEEE Access 7:85286–85297

References

Zhang L, Shen Y, Li H (2014) VSI: a visual saliency-induced index for perceptual image quality assessment. IEEE Trans Image Process 23(10):4270–4281

Zhang W, Qu C, Ma L, Guan J, Huang R (2016) Learning structure of stereoscopic image for no-reference quality assessment with convolutional neural network. Pattern Recognit 59:176–187

Chapter 2
Brief Introduction on 2D Image Quality Assessment

Abstract In this chapter, a brief introduction about 2D image quality assessment is given. Firstly, some public image quality databases are introduced which provide ground-truth information for training, testing and benchmarking. Secondly, IQA performance metrics including SROCC, KROCC, PLCC and RMSE to compare the accuracy of different IQA methods are provided. Finally, the general frameworks of 2D IQA methods containing full-reference (FR), reduced-reference (RR) and no-reference (NR) are illustrated based on specific algorithms.

Keywords 2D Image Quality Assessment · Databases · Correlation coefficient

2.1 Introduction

Visual perception information is an indispensable part of our daily life. With the development of multimedia display and transmission technology, people can obtain a lot of high-definition pictures through mobile phones, laptops, tablet computers and other electrical devices at any time. However, as an important medium of carrying information, image is inevitably polluted in the process of acquisition, reproduction, compression, storage, transmission or restoration, which finally could result in quality degradation. Therefore, objective image quality assessment (IQA) has become one of the focuses of people's research.

To begin with, researchers recognized the fact that the most reliable IQA method is human subjective judgment. Since human beings are the ultimate receivers of the visual information, the results of subjective judgement are considered to be the most accurate and reliable for perceiving the quality of images. However, directly utilizing observers to make subjective judgments on image quality is time-consuming and laborious, which is difficult to apply in real-time image processing systems. Therefore, designing an objective quality assessment algorithm to correlate with the results of subjective judgement is the mainstream in the research field of IQA. Recently, towards advancing progress on objective IQA research, a large number of classic and state-of-the-art IQA algorithms have been invented. In order to evaluate the accuracy of an IQA algorithm, ground-truth information should be obtained for training, testing and benchmarking. Ground-truth can also be recognized as image quality database in the field of IQA. A classic IQA database consists of a set of reference images and its corresponding distorted images. In addition, the most important part of IQA database is the subjective quality ratings of distorted images obtained by subjective judgement. The concept of IQA database is initiated by Video Quality Experts Group to evaluate the performance of the IQA metrics. Therefore, in this chapter, we will first briefly introduce the main publicly available image quality databases in the recent years.

Generally, the performance of an IQA algorithm is analyzed and evaluated by statistical methods. Similarity analysis often refers to solving the correlation coefficient between each objective algorithmic score and subjective (differential) mean opinion score (DMOS/MOS), so as to compare the prediction accuracy of each objective IQA algorithm. The higher the correlation, the better the performance, and the lower the correlation, the worse the performance. The most commonly used correlation coefficients include Spearman Rank-Order Correlation Coefficient (SROCC), Kendall Rank-Order Correlation Coefficient (KROCC), Pearson Product-Moment Correlation Coefficient (PLCC) and Root Mean Squared Error (RMSE). In general, PLCC and RMSE are typically used to represent the prediction accuracy and consistency, while SROCC and KROCC can be regarded as a measure of the prediction monotonicity. Higher values of PLCC, SROCC and KROCC and lower values of RMSE correspond to higher performance. The more information of the four correlation coefficients will be given in the next section.

In the next part of this chapter, we will give a brief introduction about the general framework of modern IQA methods. Generally speaking, researchers always divide IQA methods into three categories, full-reference (FR), reduced-reference (RR), and no-reference (NR). As the term suggests, this classification is based on the participation of reference during IQA operation (Wang and Bovik 2006). We define an image of undistorted version as its reference image (or original image). FR IQA methods mean that viewers can obtain more information from reference images and the quality of distorted image can be calculated by comparing the local similarity between the reference and distorted images. In contrast, NR IQA methods predict the perceptual quality of a distorted image without any information assistance of original images. The implementation of RR IQA method requires the assistance of some information from the reference image. For IQA methods belonging to different classifications, their frameworks will be different to some extent. We will spend the rest of this chapter explaining each of them. On the other hand, their basic principles are exactly the same. This means that no matter which type of algorithms we are proposing, it is inevitably required to extract quality-aware features from images, then quantify the features and map the results to the final IQA scores. For three different frameworks, we will each propose a classical algorithm to illustrate.

2.1.1 Public Image Quality Databases

Image quality database is particularly significant for objective IQA methods. The development of the former plays an important role in promoting the latter. Previous IQA methods usually performs well on a specific IQA database, but achieves poor performance on other databases. However, a qualified IQA method should not be limited to a specific IQA database. In another words, it needs to have good generalization ability. Therefore, different IQA databases are urgently needed for the research of IQA. This section summarizes several well-known public databases, such as LIVE, TID2008, TID2013, MICT, IVC, A56, LAR, WIQ, DRIQ and so on. The following gives a brief introduction to these mentioned IQA databases.

The entire Laboratory for Image and Video Engineering (LIVE) database developed by the University of Texas at Austin. LIVE includes twenty-nine high-resolution and quality color reference images, which were collected from the Internet and photographic CD-ROMS and their distorted images. The distortion types include JPEG, JEPG2000, Gaussian white noise (WN), Gaussian blur (GB) and fast fading (FF), and each distortion type is accompanied by different degrees of distortion. The total number of distorted images is 779, including 175 JPEG distorted images, 169 JPEG2000 distorted images, 145 WN distorted images, 145 Blur distorted images, 145 FF distorted images. LIVE database also provides DMOS value for each image as subjective quality score, where the higher value demonstrates the worse quality of the image.

Categorical Subjective image quality database includes 30 reference images and 866 corresponding distorted images. The distortion types include JPEG, JP2K, Additive Gaussian white noise, Additive pink Gaussian noise, Gaussian blur and Global contrast decrements (Larson and Chandler 2010).

Tampere Image Quality (TID 2008) database is created by 25 reference images from the Tampere University of Technology, Finland. The 25 reference images are obtained from the Kodark Lossless True Color Image Suite, among which the first 24 are natural images, and the 25th is computer-generated images. By introducing 17 distorted types, 1700 test images are generated, and its 17 distortion types are AGN (Additive Gaussian noise), ANC (Additive Gaussian noise in color components), SCN (Spatially correlated noise), MN (Masked noise), HFN(High frequency noise), IN (Impulse noise), QN (Quantization noise), GB (Gaussian blur), DEN (Denoising), JPEG (JPEG compression), JP2K (JP2K compression), JGTE (JPEG transmission errors), J2TE (JP2K transmission errors), NEPN (Non eccentricity pattern noise), Block (block-wise distortions of different intensity), MS (Mean shift) and CTC (Contrast change).

TID 2013 database (Ponomarenko et al. 2013) includes 3000 distorted images, and adds 7 types of distortion based on the TID 2008, which are CCS (Change of color saturation), MGN (Multiplicative Gaussian noise), CN (Comfort noise), LCNI (Lossy compression of noisy images), ICQD (Image color quantization with dither), CHA (Chromatic aberrations) and SSR (Sparse sampling and reconstruction). The value of MOS is provided in TID 2008 and TID 2013 as subjective score. The higher the score, the better the image quality.

Toyama Image Quality (MICT) database (Horita et al. 2000) presented by the Media Information and Communication Technology Laboratory of the University of Toyama, Japan, includes 14 reference images and 168 distorted images degraded by two types of distortions (JPEG and JPEG2000).

IVC database (Ninassi et al. 2006) presented by the Institute de Recherché en Communications et Cybernétique de Nantes (IRCCyN), France, contains 10 reference images and 185 distorted images degraded by five types of distortions: JPEG, JPEG2000, locally adaptive resolution (LAR), Gaussian blurring and jpeg_lumichr_r.

A57 database (Chandler and Hemami 2007) has been established by the Cornell University, USA, includes 3 reference images and 54 distorted images. The distortion types are JPEG, JPEG2000, uniform quantization of the LH subbands after a 5-level DWT (discrete wavelet transform), Gaussian blurring, Gaussian white noise and custom JPEG2000 compression by the dynamic contrast-base quantization method.

Wireless Image Quality (WIQ) database (Engelke et al. 2010) presented by the Radio Communication Group at the Blekinge Institute of Technology, Sweden, includes 7 reference images and 80 distorted images degraded by five types of distortions.

For the convenience of comparison, we list some specific information of main publicly available IQA databases in Table 2.1, including the type of images, release year of databases, the number of total images, reference images and distorted images, resolution of images and test subjects.

2.1.2 IQA Algorithm Performance Metrics

Linear correlation is the simplest way to carry out similarity analysis. Note that, linear correlation computation needs to satisfy the hypothesis that the two input sets of data must be perfectly linear correlated. However, for the objective IQA method, its output is difficult to meet this assumption. Therefore, for the sake of fairness, (Sheikh et al. 2006) suggest employing a non-linear mapping for the objective scores before computing linear correlation, that can be defined as:

$$q' = \beta_1 \left(\frac{1}{2} - \frac{1}{1 + \exp(\beta_2 (q - \beta_3))} \right) + \beta_4 q + \beta_5 \quad (2.1)$$

where q and q' are the objective scores before and after the mapping, respectively. β_1 to β_5 represent constants in this function.

As described before, four correlation coefficients including PLCC, SROCC, KROCC and RMSE are utilized for evaluating the performance of the designed objective IQA methods. In the following, the four correlation coefficients are introduced in details.

2.1 Introduction

Table 2.1 The majority of publicly available subjective image quality databases

Databases	Type	Year	Images number	Reference images	Distorted images	Distorted types	Resolution	Subjects
IRCCyN/IVC	Color	2005	195	10	185	5	512 × 512	15
LIVE	Color	2006	808	29	779	5	Various	29
A57	Gray	2007	57	3	54	6	512 × 512	7
TID2008	Color	2008	1725	25	1700	17	384 × 512	838
MICT	Color	2008	196	14	168	2	768 × 512	16
IRCCyN/MICT	Color	2008	196	14	168	2	768 × 512	24
WID/Enrico	Gray	2007	105	5	100	10	512 × 512	16
WID/BA	Gray	2009	130	10	120	2	512 × 512	17
WID/FSB	Gray	2009	215	5	210	6	512 × 512	7
WID/MW	Gray	2009	132	12	120	2	512 × 512	14
WIQ	Gray	2009	87	7	80	1	512 × 512	30
CSIQ	Color	2010	896	30	866	6	512 × 512	35
IRCCyN/DIBR	Color	2011	96	3	96	3	1024 × 768	43
HTI	Color	2011	72	12	60	1	512 × 768	18
IBBI	Color	2011	72	12	60	1	321 × 481	18
DRIQ	Color	2012	104	26	78 enhanced	3 enhancements	512 × 512	9
TID2013	Color	2013	3025	25	3000	24	512 × 512	985

(1) The Spearman's Rank-Order Correlation Coefficient (SROCC)

SROCC is the nonparametric version of the Pearson product-moment correlation. SROCC is often signified by ρ or r_s and used for measuring the strength and direction of monotonic association between two ranked variables (Bonett and Wright 2000).

There are two methods to calculate SROCC according to different conditions. If your data does not have tied ranks, the formula can be expressed as follows:

$$\rho = 1 - \frac{6 \sum d_i^2}{n(n^2 - 1)} \qquad (2.2)$$

where d_i denotes the difference in paired ranks and n denotes the number of cases.

If your data has tied ranks, the formula can be expressed as follows:

$$\rho = \frac{\sum_{i=1}^{n} (x_i - \bar{x})(y_i - \bar{y})}{\sqrt{\sum_{i=1}^{n} (x_i - \bar{x})^2 \sum_{i=1}^{n} (y_i - \bar{y})^2}} \qquad (2.3)$$

where x_i and y_i denote the ith data from x group and y group, $\bar{x}$ and $\bar{y}$ denote the average values of x group and y group.

If the value of SROCC between two variables is high, it indicates that they have a high rank correlation.

(2) The Kendall Rank-Order Correlation Coefficient (KROCC)

KROCC is a statistical indicator used for measuring the relationship between two ranked variables. Generally, the value of KROCC will be high when two variables have a similar rank. (i.e. identical for a correlation of 1) and low when two variables have a dissimilar rank. (i.e. totally different for a correlation of −1). KROCC is often signified by τ.

Suppose $(x_1, y_1), (x_2, y_2), \ldots, (x_n, y_n)$ be a set of the joint random variables X and Y respectively. Now we select two data pairs randomly from these n data pairs to form $[(x_i, y_i), (x_j, y_j)]$, where $i \neq j$. We can get a total of $\frac{n(n+1)}{2}$ data pairs. If $x_i > x_j$ and $y_i > y_j$, or $x_i < x_j$ and $y_i < y_j$, the pair is said to be concordant. If $x_i > x_j$ and $y_i < y_j$, or $x_i < x_j$ and $y_i > y_j$, the pair is said to be discordant. We count the number of concordant and discordant as P and Q respectively. If $x_i = x_j$ and $y_i > y_j$, or $x_i = x_j$ and $y_i < y_j$, the number of these pairs is X_0. If $x_i > x_j$ and $y_i = y_j$, or $x_i < x_j$ and $y_i = y_j$, the number of these pairs is Y_0. The formula can be expressed as follows:

$$\tau = \frac{P - Q}{\sqrt{(P + Q + X_0)(P + Q + Y_0)}} \qquad (2.4)$$

(3) Pearson Product-Moment Correlation Coefficient (PLCC)

PLCC is the covariance of the two variables divided by the product of their standard deviations. PLCC is used to evaluate the linear correlation between

2.1 Introduction

two groups of data. If the absolute value of PLCC is closer to 1, it indicates that the relationship between two variables can be expressed by a linear equation. The formula can be expressed as follows:

$$PLCC = \frac{Cov(X,Y)}{\sigma_X \sigma_Y} \quad (2.5)$$

(4) Root Mean Square Error (RMSE)

RMSE represents the sample standard deviation of the differences between predicted values and observed values. RMSE is a measure of accuracy, to compare forecasting errors of different models for a particular dataset and not between datasets, as it is scale-dependent (Hyndman and Koehler 2006).

RMSE is always non-negative, and a value of 0 (almost never implemented in practice) indicates a complete match with the data. Generally speaking, a lower RMSE is better than a higher one. The formula can be expressed as follows:

$$RMSE = \sqrt{\frac{\sum_{i=1}^{n}(x_i - y_i)^2}{n}} \quad (2.6)$$

2.1.3 Typical Objective IQA Algorithms

2.1.3.1 The Framework of Full-Reference Methods and Typical Algorithms

Objective IQA algorithms can be classified according to the availability of a reference image, with which the distorted image is to be compared. Full-reference methods mean that the information of a complete reference image is assumed to be known. The earliest and simplest full-reference quality metric is the MSE (mean squared error). The function of MSE can be expressed as follows:

$$MSE = \frac{1}{N} \sum_{i=1}^{N} (T(i) - R(i))^2 \quad (2.7)$$

where T and R denote the gray-scale images of the tested and reference images respectively, and N is the total pixel number of the tested image. Obviously, MSE can be used as a metric to measure the difference between the distorted images and its reference images pixel by pixel, but the pixel difference is not the perceptual quality difference for images, which is why the accuracy of MSE is widely criticized (Wang and Bovik 2009). In order to improve the prediction accuracy of image quality, a more former and advanced expression is proposed, which extracts quality-aware features from images such as luminance, information content, local energy, local texture, gradient and so on. All of these features can be considered to be sensitive to the image quality. Then, the dissimilarity measurement is operated on the quality-aware features instead of raw images. And the final step is to map all of these results into an objective quality score. The framework of FR IQA methods can be summarized in Fig. 2.1.

As a typical example, the method of structural similarity (SSIM) compares the local patterns of pixel intensities that have been normalized for luminance and contrast (Wang et al. 2004). SSIM takes advantage of the high similarity of natural image signals. These signals' pixels exhibit strong dependencies, especially when they are spatially proximate. Meanwhile, these dependencies carry important information about the structure of the objects in the visual scene. The paradigm of SSIM is a top-down approach to simulate the hypothesized functionality of the overall HVS (human visual system).

As Fig. 2.2 shown above, x and y are used as two nonnegative input image signals, which have been aligned with each other (e.g., spatial patches extracted from each image). Generally, one of the signals is considered as reference without distortion, and then this system will serve as a quantitative measurement of the quality of the second signal. As seen clearly in Fig. 2.2, this system compares luminance, contrast and structure respectively. First, the luminance of each signal is compared. The mean intensity is defined as follows (for discrete signals):

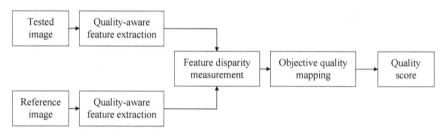

Fig. 2.1 Typical framework of FR methods

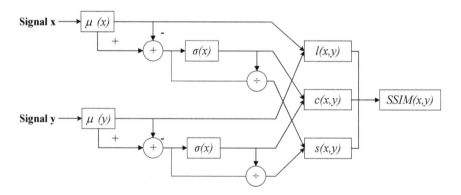

Fig. 2.2 Diagram of the structural similarity (SSIM) measurement system

2.1 Introduction

$$\mu_x = \frac{1}{N}\sum_{i=1}^{N} x_i \tag{2.8}$$

Second, the signal contrast is estimated by the standard deviation (the square root of variance). An unbiased estimate in discrete form can be expressed by

$$\sigma_x = \left(\frac{1}{N-1}\sum_{i=1}^{N}(x_i - u_x)^2\right)^{1/2} \tag{2.9}$$

Third, the structure comparison is conducted on normalized signals. Then the expression of SSIM can be written as:

$$S(x, y) = f(l(x, y), c(x, y), s(x, y)) \tag{2.10}$$

where $l(x, y)$, $c(x, y)$, $s(x, y)$ are relatively independent.

The similarity measure should satisfy the following conditions:

(1) Symmetry: $S(x, y) = S(y, x)$.
(2) Boundedness: $S(x, y) \leq 1$.
(3) Unique maximum: $S(x, y) = 1$ if and only if $x = y$.

In order to realize the above three conditions, we define:

$$l(x, y) = \frac{2\mu_x\mu_y + C_1}{\mu_x^2 + \mu_y^2 + C_1} \tag{2.11}$$

where the constant C_1 is included to avoid instability when $\mu_x^2 + \mu_y^2$ is very close to zero. Specifically, we choose

$$C_1 = (K_1 L)^2 \tag{2.12}$$

where $K_1 \ll 1$ is a small constant, and L is the dynamic range of pixel values (255 for 8-bit grayscale images). Equation 2.10 is also in accordance with Weber's law, which has been widely used to model light adaptation (also known as luminance masking) in the HVS.

The equation form of the contrast comparison function is similar, which can be expressed as follows:

$$c(x, y) = \frac{2\sigma_x\sigma_y + C_2}{\sigma_x^2 + \sigma_y^2 + C_2} \tag{2.13}$$

where $C_2 = (K_2 L)^2$, and the constant $K_2 \ll 1$. An important feature of this function is that this measure is less sensitive to the case of high base contrast than low base contrast. This feature is exactly consistent with the contrast-masking feature of the HVS.

The function of structure comparison is defined as:

$$s(x, y) = \frac{\sigma_{xy} + C_3}{\sigma_x \sigma_y + C_3} \tag{2.14}$$

where C_3 is a small constant in both denominator and numerator. σ_{xy} is the covariance of x and y, which can be calculated as follows:

$$\sigma_{xy} = \frac{1}{N-1} \sum_{i=1}^{N} (x_i - \mu_x)(y_i - \mu_y) \tag{2.15}$$

After getting these three equations, the resulting similarity measure between signals x and y can be computed as:

$$SSIM(x, y) = [l(x, y)]^\alpha \cdot [c(x, y)]^\beta \cdot [s(x, y)]^\gamma \tag{2.16}$$

where $\alpha > 0$, $\beta > 0$ and $\gamma > 0$ are parameters used to adjust the weight of the three components. It is obviously that this equation satisfies the three conditions given above. To simplify the expression, set $\alpha = \beta = \gamma = 1$ and $C_3 = C_2/2$. And the new form of expression can be written as follows:

$$SSIM(x, y) = \frac{(2\mu_x \mu_y + C_1)(2\sigma_{xy} + C_2)}{(\mu_x^2 + \mu_y^2 + C_1)(\sigma_x^2 + \sigma_y^2 + C_2)} \tag{2.17}$$

It is useful to apply the SSIM index locally rather than globally in the field of IQA research. The main reason is that only a local area in the image can be perceived with high resolution by the human observers at one time instance at specific viewing distances (Wang and Bovik 2001). Therefore, the local statistics μ_x, σ_x and σ_{xy} are computed within a local 8×8 square window, which moves pixel-by-pixel over the entire image. Each step of calculating the local statistics and SSIM index is based on the pixels within the local window. Finally, a SSIM index mapping matrix is obtained, which is composed of local SSIM index. One problem with this approach is that the generated SSIM index map often shows undesirable "blocking" artifacts. To solve this problem, an 11×11 circular-symmetric Gaussian weighting function is used with standard deviation of 1.5 samples. The local statistics are modified accordingly as:

$$\mu_x = \sum_{i=1}^{N} w_i x_i \tag{2.18}$$

$$\sigma_x = \left(\sum_{i=1}^{N} w_i (x_i - u_x)^2 \right)^{\frac{1}{2}} \tag{2.19}$$

2.1 Introduction

$$\sigma_{xy} = \sum_{i=1}^{N} w_i(x_i - \mu_x)(y_i - \mu_y) \tag{2.20}$$

The quality maps exhibit a locally isotropic property with such a windowing approach. After some experiments, setting $K_1 = 0.01$ and $K_2 = 0.03$ to improve the performance of SSIM index. In practice, a mean SSIM index is used to evaluate the overall image quality.

$$MSSIM(X, Y) = \frac{1}{M} \sum_{j=1}^{M} SSIM(x_j, y_j) \tag{2.21}$$

where X and Y are the reference and distorted images, respectively; x_j and y_j denote the image contents at the jth local window; M is the total number of local windows of the image. Using the average value of the SSIM index of overall images is the basic method. The different weighting functions can be introduced into the SSIM method in different application (Privitera and Stark 2000; Rajashekar et al. 2003).

SSIM is based on the hypothesis that HVS is adapted to extract the structural information. In this case, the similarity of structure information plays a role in judging image quality. Experimental results show that the SSIM index achieves better performance than early IQA metrics based on simulating HVS, but SSIM still needs further improvement and subsequent research.

2.1.3.2 The Framework of Reduced-Reference Methods and Typical Algorithms

Reduced-reference (RR) image quality assessment is a relatively new research topic compared with the FR and NR methods. RR IQA method provides a feasible solution that delivers a trade-off between FR and NR. The advantages and disadvantages of FR and NR are obvious. FR methods make full use of the reference image information, which usually achieve the highest accuracy. But it's hard to obtain all the information about the reference. On the contrary, NR methods are practically applicable in almost all situations, but the predicting accuracy is relatively low. RR methods are designed to predict perceptual image quality with only partial information about the reference images. Therefore, RR methods can achieve better performance than NR, and do not have too strict requirements as FR (Li and Wang 2009).

RR methods are highly scenario-dependent. In different application scenarios, the available reference information might vary, so different methods should be applied. Some desirable properties of RR features include: (1) they should provide an efficient summary of the reference image, (2) they should be sensitive to a variety of image distortions, (3) they should have good perceptual relevance.

A typical framework of RR methods can be shown as Fig. 2.3.

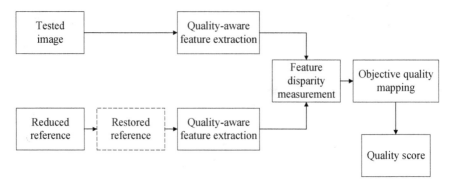

Fig. 2.3 Typical framework of RR methods

Compared with the framework in Fig. 2.1, the only explicit modification in Fig. 2.3 is the additive processing of the restored reference, for different RR methods should be designed for different applications. RR methods are divided into many different types correspondingly, and the classification itself is quite complex. How to select proper situations that make RR methods maximally useful is a challenging problem. And we will introduce a typical RR method called wavelet domain RR measure. Figure 2.4 shows how this RR quality analysis system be deployed.

The system includes a feature extraction process at the sender side and a quality analysis process (also have a feature extraction process) at the receiver side. The extracted RR features often have a lower data rate than the image data and are typically transmitted to the receiver through an ancillary channel.

For this method, firstly suppose $p(x)$ and $q(x)$ denote the probability density functions of the wavelet coefficients in the same subband of two images respectively. Suppose $x = \{x_1, \ldots, x_N\}$ be a set of N randomly and independently selected coefficients. The log-likelihoods of x from $p(x)$ and $q(x)$ can be expressed as:

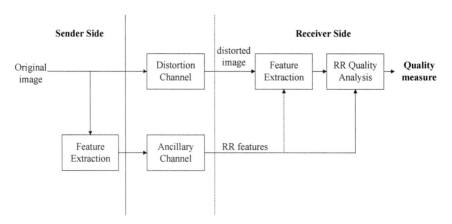

Fig. 2.4 Deployment of RR IQA system

2.1 Introduction

$$l(p|x) = \frac{1}{N} \sum_{n=1}^{N} \log p(x_n) \qquad (2.22)$$

$$l(q|x) = \frac{1}{N} \sum_{n=1}^{N} \log q(x_n) \qquad (2.23)$$

The difference of the log-likelihood between $p(x)$ and $q(x)$ is

$$l(p|x) - l(q|x) = \frac{1}{N} \sum_{n=1}^{N} \log \frac{p(x_n)}{q(x_n)} \qquad (2.24)$$

Assume that $p(x)$ is the true probability density distribution of the coefficients, when N is large, the difference of the log-likelihoods between $p(x)$ and $q(x)$ asymptotically approaches the Kullback-Leibler distance (KLD) between p $p(x)$ and $q(x)$:

$$l(p|x) - l(q|x) \to d(p\|q) = \int p(x) \log \frac{p(x)}{q(x)} dx \qquad (2.25)$$

KLD is used to quantify the difference of wavelet coefficient distributions between reference and distorted images. To estimate the KLD between them, the coefficient histograms of both the reference and distorted images should be obtained. The latter can be easily computed from the received distorted image, but the former is really difficult to realize. The marginal distribution of the coefficients in individual wavelet subbands can be well-fitted with a 2-parameter generalized Gaussian density model (Moulin and Liu 1999; Mallat 1989):

$$p_m(x) = \frac{\beta}{2\alpha \Gamma(1/\beta)} e^{-(|x|/\alpha)^\beta} \qquad (2.26)$$

where $\Gamma(\alpha) = \int_0^\infty t^{\alpha-1} e^{-t} dt$ (a > 0) is the Gamma function. This model provides a very efficient means to summarize the coefficient histogram of the reference. In addition to two parameters α and β, the prediction error is added as a third RR feature parameter, which can be defined as follows:

$$d(p_m\|p) = \int p_m(x) \log \frac{p_m(x)}{p(x)} dx \qquad (2.27)$$

In practice, this quantity should be evaluated numerically using histograms:

$$d(p_m\|p) = \sum_{i=1}^{L} P_m(i) \log \frac{P_m(i)}{P(i)} \qquad (2.28)$$

where $P(i)$ and $P_m(i)$ are the normalized heights of the i-th histogram bins, and L represents the number of bins in the histograms.

At the receiver side, the KLD between $p_m(x)$ and $q(x)$ should be computed:

$$d(p_m \| q) = \int p_m(x) \log \frac{p_m(x)}{q(x)} dx \qquad (2.29)$$

Because a distorted image may not be a natural image anymore and may not be well-fitted with a GGD model, the histogram bins of the wavelet coefficients is extracted as the feature at the receiver side. The KLD between $p(x)$ and $q(x)$ can be estimated as

$$\hat{d}(p \| q) = d(p_m \| q) - d(p_m \| p) \qquad (2.30)$$

The estimation error is

$$d(p \| q) - \hat{d}(p \| q) = \int [p(x) - p_m(x)] \log \frac{p(x)}{q(x)} dx \qquad (2.31)$$

Finally, the overall distortion between the distorted and reference images is:

$$D = \log_2 \left(1 + \frac{1}{D_0} \sum_{k=1}^{K} \left| \hat{d}^k(p^k \| q^k) \right| \right) \qquad (2.32)$$

where K is the number of subbands, p^k and q^k denote the probability density functions of the k-th subbands in the reference and distorted images respectively, $\hat{d}^k$ is the estimation of the KLD between p^k and q^k, and D_0 is a constant to control the scale of the distortion measure.

The feature extraction system for the reference image at the sender side and the RR quality analysis system at the receiver side are illustrated in Figs. 2.5 and 2.6.

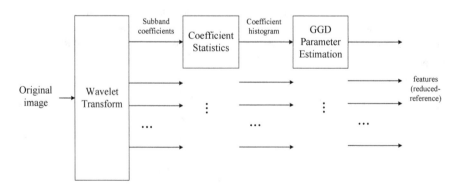

Fig. 2.5 Feature extraction system at the sender side

2.1 Introduction

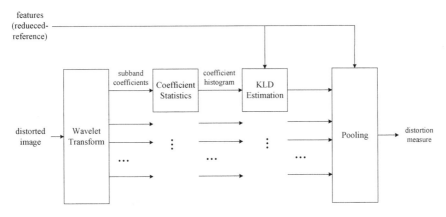

Fig. 2.6 Quality analysis system at the receiver side

This method is of great significance for practical use for the following reasons: (1) it is a general-purpose method applicable to a wide range of distortion types, (2) it has a relatively low RR data rate, (3) the method is easy to implement, computationally efficient, and uses only a few parameters, (4) the method is insensitive to small geometric distortions such as spatial translation, rotation, and scaling.

2.1.3.3 The Framework of No-Reference Methods and Typical Algorithms

The concept of NR (is also called blind) methods is as its name implies. The NR method must evaluate the quality of a real-world image without any knowledge about original reference information from the high-quality image. This seems like an impossible task on the surface. However, without looking at the original image, humans also can easily identify high-quality images versus low-quality images. It's obvious that human visual system is making use of a very substantial and effective pool of information about images in making subjective judgments of image quality. Therefore, quality-aware features can still be obtained in the process of NR methods, and the challenging problem is how to build the connection between quality-aware features and quality indices in the absence of references.

There are generally two kinds of approaches that are commonly adopted in recent research. The first kind is to build artificial original images. This kind of methods is divided into two sub-categories. The former constructs an ideal method and regard the difference between a test image and artificial reference as its quality score (Moorthy and Bovik 2011; Saad et al. 2012). These methods assume that the original natural images share similar statistics (Simoncelli and Olshausen 2001). The typical framework is shown in Fig. 2.7. The latter use de-noising schemes to recover the corresponding reference for each distorted image (Portilla et al. 2003). First a classifier is set to predict the distortion types, because the de-noising algorithms for different distortion types may be different (Moorthy and Bovik 2010). In the early stage of NR IQA research, many scholars have proposed lots of methods for specific distortion (Wu and Yuen 1997; Wang et al. 2000, 2002; Yu et al. 2002; Sheikh et al. 2002, 2005; Marziliano et al. 2002). The NR methods for general purposes are only largely emerging after around 2010s.

The second kind of NR IQA methods emerge more recently, which directly maps the quality-aware features to the final quality scores without constructing an artificial original image. Obviously, in addition to the consistency between the features and human subjective perception, the performance of such approach also depends on the quality of the mapping functions and the training databases to a large extent. However, due to the omission of the dissimilarity measurement, these methods do not rely on proper distance quantification schemes, and are usually computationally faster. The second methods are becoming more and more popular in practical application. But we are not sure which kind of methods are better, it depends on the specific application. The framework of the second kind of methods is as follows (Fig. 2.8).

We will first introduce a NR IQA method based on the first kind framework, which attempts to develop a general-purpose NR scheme by employing topographic independent components analysis (TICA) to extract completely independent features. TICA can be regarded as a generalization of independent subspace analysis model to imitate the topographic representation in the visual cortex. The essential preprocessing steps include (i) images are divided into several patches of 16×16 pixels to form 256 dimensional vectors; (ii) the mean greyscale value is removed from each image patch. For each patch, 243 principal components with the largest variances are retained by the means of dimension reducing, and then the data are whitened by

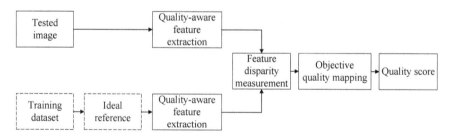

Fig. 2.7 Typical framework of NR methods by constructing an ideal reference

2.1 Introduction

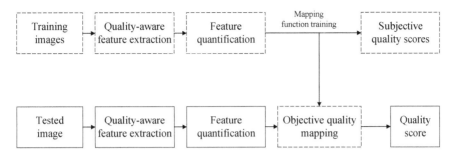

Fig. 2.8 Typical framework of NR methods by training quality mapping function

normalizing the variances of these principal components

$$T_i = \sum_{x,y} w_i(x, y) z(x, y) \qquad (2.33)$$

where w_i is the independent component weight which is orthonormalized, T_i is the one piece of the output (e.g. features of the image) and z is the whitened data.

The statistically independent features can be calculated by

$$F_i = \sqrt{\sum_j^n h(i, j) T_j^2} \qquad (2.34)$$

where $h(i, j)$ represents the proximity between the i-th and j-th components.

A generalized Gaussian distribution (GGD) model is used to characterize the independent components F_i for subband responses of natural scenes tend to follow a non-Gaussian distribution. A lot of experiments have been conducted to prove that the distortions of an image will affect the probability density function. Thus, an ideal feature distribution can be obtained through the training procedure. Once trained by a set of pristine images, it is reasonable that the ideal distribution can serves as a 'reference' for assessing the quality of distorted images.

Therefore, the quality of the distorted image can be evaluated by quantifying the variation between the ideal distribution and the actual distribution approximated from the histogram of the distorted image. This method uses Kullback-Leibler distance (KLD) to measure the distances of all the features

$$D(p(x)\|q(x)) = \sum_{i=1}^{N} r_i \times \int p_i(x) \log \frac{p_i(x)}{q_i(x)} dx \qquad (2.35)$$

where N is the total number of independent features, r_i represents a factor to adjust the weight of different features, and $p(x)$ and $q(x)$ denote the ideal and actual probability density functions, respectively.

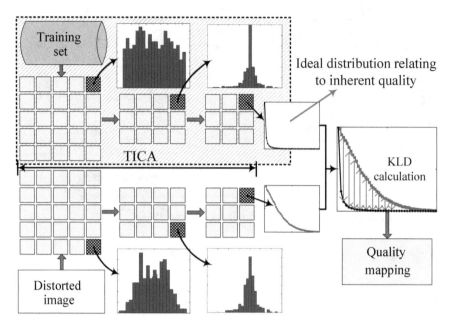

Fig. 2.9 NR scheme with topographic independent components analysis

Finally, D is mapped with a five-parameter monotonic logistic function to produce an objective score. The overall scheme of the proposed method is illustrated in Fig. 2.9.

Then we will introduce a BRISQUE (blind/no-reference image spatial quality evaluator) model based on the second framework. It uses scene statistics of locally normalized luminance coefficients to quantify possible losses of "naturalness" in the image due to the presence of distortions. The specific steps of this algorithm can be summarized as follows. Firstly, locally normalized luminance information is calculated via the method of local mean subtraction and divisive normalization. The procedure of local normalization can be concluded as following:

$$\hat{I}(i,j) = \frac{I(i,j) - \mu(i,j)}{\sigma(i,j) + C} \tag{2.36}$$

where $i \in 1, 2, \ldots, M$, $j \in 1, 2, \ldots, N$ represent the spatial indices, M and N are the image height and width respectively, $C = 1$ is a constant to prevent instabilities from happening when the denominator tends to zero. $\mu(i,j)$ and $\sigma(i,j)$ can be expressed as

$$\mu(i,j) = \sum_{k=-K}^{K} \sum_{l=-L}^{L} w_{k,l} I_{k,l}(i,j) \tag{2.37}$$

2.1 Introduction

$$\sigma(i, j) = \sqrt{\sum_{k=-K}^{K} \sum_{l=-L}^{L} w_{k,l}(I_{k,l}(i, j) - u(i, j))^2} \quad (2.38)$$

where $w = \{w_{k,l} | k = -K, ..., K; l = -L, ..., L\}$ denotes a 2D circularly-symmetric Gaussian weighting function sampled out to 3 standard deviations and rescaled to unit volume, and $K = L = 3$.

Then the pre-processing model (2.36) is utilized to calculate the transformed luminance $\hat{I}(i, j)$ as mean subtracted contrast normalized (MSCN) coefficients. And the MSCN coefficients have characteristic statistical properties that are changed by the presence of distortion, and that quantifying these changes will make it possible to predict the type of distortion affecting an image as well as its perceptual quality. The visualization of this property can be seen in Fig. 2.10.

Generalized Gaussian distribution (GGD) can be utilized to effectively capture a broader spectrum of distorted image statistics. Then the zero mean distribution is chosen to fit the MSCN empirical distributions from distorted images as well as undistorted ones. The model is given by:

$$f(x; \alpha, \sigma^2) = \frac{\alpha}{2\beta\Gamma(1/\alpha)} \exp\left(-\left(\frac{|x|}{\beta}\right)^\alpha\right) \quad (2.39)$$

where

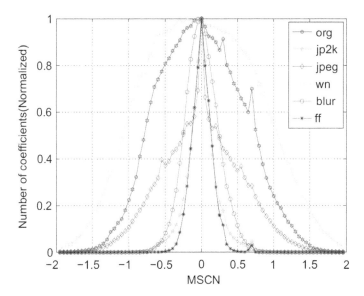

Fig. 2.10 Histogram of MSCN coefficients for a natural undistorted image and its various distorted versions. Distortions from the LIVE IQA database. jp2k: JPEG2000. jpeg: JPEG compression. WN: additive white Gaussian noise. blur: Gaussian blur. ff: Rayleigh fast-fading channel simulation

$$\beta = \sigma \sqrt{\frac{\Gamma(1/\alpha)}{\Gamma(3/\alpha)}} \tag{2.40}$$

and $\Gamma(\cdot)$ is the gamma function:

$$\Gamma(\alpha) = \int_0^\infty t^{\alpha-1} e^{-t} dt \quad \alpha > 0 \tag{2.41}$$

For each image, 2 parameters (α, σ^2) can be estimated from a GGD fit of the MSCN coefficients.

The statistical relationship between neighboring pixels can be computed by

$$H(i, j) = \hat{I}(i, j)\hat{I}(i, j+1) \tag{2.42}$$

$$V(i, j) = \hat{I}(i, j)\hat{I}(i+1, j) \tag{2.43}$$

$$D1(i, j) = \hat{I}(i, j)\hat{I}(i+1, j+1) \tag{2.44}$$

$$D2(i, j) = \hat{I}(i, j)\hat{I}(i+1, j-1) \tag{2.45}$$

where horizontal (H), vertical (V), main-diagonal (D1) and secondary-diagonal (D2) are the empirical distributions of pairwise products of neighboring MSCN coefficients along four orientations. Then the asymmetric generalized Gaussian distribution (AGGD) model is utilized to fit pairwise products of neighboring MSCN coefficients. The AGGD with zero mode is given by:

$$f(x; \nu, \sigma_l^2, \sigma_r^2) = \begin{cases} \frac{\nu}{\beta_l + \beta_r \Gamma(\frac{1}{\nu})} \exp\left(-\left(\frac{-x}{\beta_l}\right)^\nu\right) & x < 0 \\ \frac{\nu}{\beta_l + \beta_r \Gamma(\frac{1}{\nu})} \exp\left(-\left(\frac{x}{\beta_r}\right)^\nu\right) & x \geq 0 \end{cases} \tag{2.46}$$

where

$$\beta_l = \sigma_l \sqrt{\frac{\Gamma(\frac{1}{\nu})}{\Gamma(\frac{3}{\nu})}} \tag{2.47}$$

$$\beta_r = \sigma_r \sqrt{\frac{\Gamma(\frac{1}{\nu})}{\Gamma(\frac{3}{\nu})}} \tag{2.48}$$

The shape parameter ν controls the 'shape' of the distribution while σ_l^2 and σ_r^2 are scale parameters that control the spread on each side of the mode respectively. The parameters $(\eta, \nu, \sigma_l^2, \sigma_r^2)$ of the best AGGD fit are extracted where η is given by:

2.1 Introduction

Table 2.2 Summary of features extracted in order classify and quantify distortions

Feature ID	Feature description	Computation procedure
f_1–f_2	Shape and variance	Fit GGD to MSCN coefficients
f_3–f_6	Shape, mean, left variance, right variance	Fit AGGD to H pairwise products
f_7–f_{10}	Shape, mean, left variance, right variance	Fit AGGD to V pairwise products
f_{11}–f_{14}	Shape, mean, left variance, right variance	Fit AGGD to D1 pairwise products
f_{15}–f_{18}	Shape, mean, left variance, right variance	Fit AGGD to D2 pairwise products

$$\eta = (\beta_r - \beta_l) \frac{\Gamma\left(\frac{2}{v}\right)}{\Gamma\left(\frac{1}{v}\right)} \qquad (2.49)$$

Thus, a total of 36 features (18 at each scale) are used to identify distortions and to perform distortion-specific quality assessment, see Table 2.2 for details.

Finally, a procedure of mapping is learned from feature space to quality scores using a regression module, yielding a measure of image quality. There the LIBSVM package (Chang and Lin 2001) is utilized to implement the support vector machine regressor (SVR) (Schölkopf et al. 2000) with a radial basis function (RBF) kernel.

2.2 Summary

This chapter systematically discusses some related knowledge about 2D IQA works. Databases can be regarded as one of the most crucial and essential components for training, testing and benchmarking of a new algorithms. Thus we first introduce some most frequently used 2D image quality databases. With the development of stereoscopic and autostereoscopic displays (3D) technologies (Lambooij et al. 2011; Seuntiens et al. 2006), the perceived depth information is enhanced. Several 3D image quality databases have been proposed at present. However, for the proposed 3D image quality databases, the number of images and distortion types are too few to show the reliable of the databases. This brings a lot of difficulties to the 3D IQA works. Then we give a brief introduction about four typical performance metrics of image quality assessment.

The main content of this chapter can be concluded by three kind of frameworks of IQA methods, which are full-reference (FR), reduced-reference (RR), and no-reference (NR) methods. For each method, we explained the mechanism in detail and given the frameworks of some typical related algorithms. Although NR methods are generally exhibiting comparatively worse performance than FR methods, even in the current stage, its value in practical application is higher, which is our main research direction at present.

References

Bonett DG, Wright TA (2000) Sample size requirements for Pearson, Kendall, and Spearman correlations. Psychometrika 65:23–28

Chandler DM, Hemami SS (2007) A57 database. [EB/OL]. Available: http://foulard.ece.cornell.edu/dmc27/vsnr.vsnr.html

Chang CC, Lin CJ (2001) LIBSVM: a library for support vector machines. [Online]. Available: https://www.csie.ntu.edu.tw/~cjlin/libsvm/

Engelke U, Zepernick HJ, Kusuma M (2010) Wireless imaging quality database. [EB/OL]. Available: http://www.bth.se/tek/rcg.nsf/pages/wiq-db

Horita Y, Shibata K, Kawayoke Y (2000) MICT image quality evaluation database. [EB/OL]. Available: http://mict.eng.u-toyama.ac.jp/mictdb.html

Hyndman RJ, Koehler AB (2006) Another look at measures of forecast accuracy. Int J Forecast 22(4):679–688

Lambooij M, IJsselsteijn W, Bouwhuis DG, Heynderickx I (2011) Evaluation of stereoscopic images: beyond 2D quality. IEEE Trans Broadcast 57(2):432–444

Larson EC, Chandler DM (2010) Most apparent distortion: full-reference image quality assessment and the role of strategy. J Electron Imaging 19(1):011006

Li Q, Wang Z (2009) Reduced-reference image quality assessment using divisive normalization-based image representation. IEEE J Sel Topics Signal Process 3(2):202–211

Mallat SG (1989) Multifrequency channel decomposition of images and wavelet models. IEEE Trans Acoust Speech Signal Process 37(12):2091–2110

Marziliano P, Dufaux F, Winkler S, Chen T (2002) Perceptual blur and ringing metrics: application to JPEG2000. Signal Process: Image Commun 19(2):163–172

Moorthy AK, Bovik AC (2010) A two-step framework for constructing blind image quality indices. IEEE Signal Process Lett 17(5):513–516

Moorthy AK, Bovik AC (2011) Blind image quality assessment: from natural scene statistics to perceptual quality. IEEE Trans Image Process 20(12):3350–3364

Moulin P, Liu J (1999) Analysis of multiresolution image denoising schemes using a generalized Gaussian and complexity priors. IEEE Trans Inf Theory 45(3):909–919

Ninassi A, Calet PL, Autrusseau F (2006) Pseudo no reference image quality metric using perceptual data hiding. Proc SPIE Hum Vis Electron Imaging 6057:146–157

Ponomarenko NN, Ieremeiev O, Lukin VV (2013) Color image database TID2013: peculiarities and preliminary results. In: European workshop on visual information processing, pp 106–111

Portilla J, Strela V, Wainwright MJ, Simoncelli EP (2003) Image denoising using scale mixtures of Gaussians in the wavelet domain. IEEE Trans Image Process 12(11):1338–1351

Privitera CM, Stark LW (2000) Algorithms for defining visual regions-of-interest: comparison with eye fixations. IEEE Trans Pattern Anal Mach Intell 22(9):970–982

Rajashekar U, Cormack LK, Bovik AC (2003) Image features that draw fixations. In: Proceedings 2003 international conference on image processing, Barcelona, Spain, pp 313–316

Saad M, Bovik AC, Charrier C (2012) Blind image quality assessment: a natural scene statistics approach to perceptual quality. IEEE Trans Image Process 21(8):3339–3352

Schölkopf B, Smola AJ, Williamson RC, Bartlett PL (2000) New support vector algorithms. Neural Comput 12(5):1207–1245

Seuntiens P, Meesters L, IJsselsteijn W (2006) Perceived quality of compressed stereoscopic images: effects of symmetric and asymmetric jpeg coding and camera separation. ACM Trans Appl Percept 3(2):95–109

Sheikh HR, Wang Z, Cormack L, Bovik AC (2002) Blind quality assessment for JPEG2000 compressed images. In: Conference record of the thirty-sixth Asilomar conference on signals, systems and computers, Pacific Grove, CA, USA, 2, pp 1735–1739

Sheikh HR, Bovik AC, Cormack L (2005) No-reference quality assessment using natural scene statistics: JPEG2000. IEEE Trans Image Process 14(11):1918–1927

References

Sheikh HR, Sabir MF, Bovik AC (2006) A statistical evaluation of recent full reference image quality assessment algorithms. In: IEEE transactions on image processing 15(11):3440–3451. https://doi.org/10.1109/TIP.2006.881959

Simoncelli EP, Olshausen BA (2001) Natural image statistics and neural representation. Ann Rev Neurosci 24:1193–1216

Wang Z, Bovik AC (2001) Embedded foveation image coding. IEEE Trans Image Process 10(10):1397–1410

Wang Z, Bovik AC (2006) Modern image quality assessment. Morgan & Claypool, San Rafael, California, United States

Wang Z, Bovik AC (2009) Mean squared error: love it or leave it? A new look at signal fidelity measures. IEEE Signal Process Mag 26(1):98–117

Wang Z, Bovik AC, Evans BL (2000) Blind measurement of blocking artifacts in images. In: Proceedings of IEEE international conference on image processing, Vancouver, BC, Canada, 3, pp 981–984

Wang Z, Sheikh HR, Bovik AC (2002) No-reference perceptual quality assessment of JPEG compressed images. In: Proceedings of IEEE international conference on image processing, Rochester, NY, USA, pp 477–480

Wang Z, Bovik AC, Sheikh HR, Simoncelli EP (2004) Image quality assessment: from error visibility to structural similarity. IEEE Trans Image Process 13(4):600–612

Wu HR, Yuen M (1997) A generalized block-edge impairment metric for video coding. IEEE Signal Process Lett 4(11):317–320

Yu Z, Wu HR, Winkler S, Chen T (2002) Vision-model-based impairment metric to evaluate blocking artifact in digital video. Proc IEEE 90:154–169

Chapter 3
Difference Between 2D and Stereoscopic Image Quality Assessment

Abstract With the rapid development of stereo and multi-view systems and a wide adoption of these systems, stereoscopic image quality assessment (SIQA) has become an important and challenging problem faced in numerous application such as 3D films, stereo visualization and 3D enhancement. Compared with 2D image quality assessment methods, SIQA needs considering complex binocular visual properties. The difference between SIQA and 2D IQA is introduced firstly in this chapter. Then stereoscopic image quality databases are listed and discussed detailedly. Finally, the general designed frameworks of SIQA methods are provided.

Keywords Image quality assessment · Stereoscopic image · Binocular visual properties · Visual discomfort

3.1 Introduction

By far we have discussed the evolution and general frameworks of 2D IQA in the Chap. 2, where the subjective score can be obtained from individual nature images. With the development of multimedia production and display techniques, three-dimensional (3D) applications like 3D films, television, gaming and virtual reality (VR) are becoming more and more popular in human's daily life, which drives researchers to turn to the research of stereoscopic image quality assessment (SIQA) (Chen et al. 2013a; Shao et al. 2015; Ding and Zhao 2017).

The problem of SIQA is defined to predict the subjective quality score of stereoscopic images accurately. The term stereoscopic image we refer to here, or 3D image, stereopair, consists of a pair of monocular images captured by two cameras. The two monocular images contain the same contents with slightly different horizontal position, which that creates an impression of depth information. Figure 3.1 illustrates a pair of monocular views from LIVE 3D IQA database (Moorthy et al. 2013), from which we can see that the locations of a particular object in the left view are all in the right of that in the right view that is due to the existence of depth difference between the two monocular images. In addition, each object is on the same horizontal level in the two monocular views.

Fig. 3.1 The **a** left view and **b** right view of a stereoscopic image

The distortion type of stereopairs is similar to that in 2D IQA databases, mainly including JPEG compression, JPEG2000 (JP2K) compression, white noise (WN), gaussian blur (GB) and fast fading (FF). Each distortion type is symmetrically assigned to the left and right views of stereopairs with five distortion levels. Different from 2D IQA databases, there exists an asymmetrical distortion type that the distortion degrees of left and right views are different, as depicted in Fig. 3.2. The special type of image distortion could introduce a kind of binocular visual properties, among which binocular rivalry need to be considered and treated in dealing with the quality prediction problem of stereopairs. Obviously, it also is claimed that the generation of

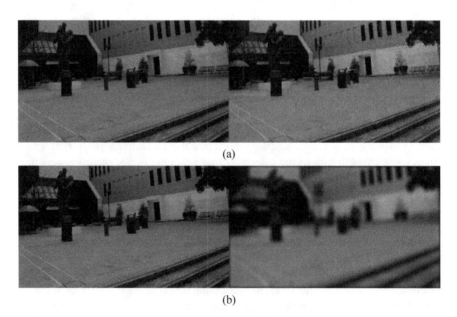

Fig. 3.2 Examples of asymmetrically distortion type from LIVE 3D IQA database, where a stereoscopic image with right view distorted by **a** JPEG compression and **b** Gaussian blur

3.1 Introduction

binocular vision is not merely a simple fusion of two views but includes the development of the depth perception, which all will be discussed in the following section in detail.

Let us now turn to explore the main difference between plane images and stereo images. Owing to unique binocular visual properties contained in stereo images, assessing the quality of 3D images is a more complicated task than its 2D counterparts. Even if considerable progress has been made in objective 2D IQA, evaluating the perceptual quality of stereoscopic images is still an unsolved and puzzled research topic. The fact has been declared by psychological and physiological research that binocular vision is not simple summation of two monocular image process. Therefore, in early SIQA works (Benoit et al. 2008b; Yasakethu et al. 2008; You et al. 2010), directly applying 2D IQA methods for SIQA often fails to achieve satisfactory results. Later research considering binocular visual properties, as expected, is capable of achieving much promising performance than before. Unfortunately, human binocular vision, denoted as a complicated visual process transmitted and analyzed by a series of simple and complex cells in human brain, cannot be modelled by simple mathematical functions effectively and accurately. The modelling of binocular effects is the main enemy facing SIQA researchers, as well are considered as very significant research topics in image processing, computer vision and artificial intelligence nowadays. In another words, it is indeed the existence of binocular effects that makes the study of SIQA more interesting and meaningful, otherwise the 2D IQA methods would be sufficient in the 3D field. Human binocular vision, including some of the appealing binocular effects having influences upon stereoscopic image quality perception, will be discussed and analyzed detailly in Sect. 3.2.

Another key information that needs to be discussed is the related technologies for 3D image quality database constructions, including image registration, subjective test, distortion type and acquisition protocols (Moorthy et al. 2013; Chen et al. 2013b). So far, there are several established SIQA databases have been released for 3D IQA, which derives the quick development of SIQA research and allows a fair comparison of SIQA metrics. It is worthy to notice that there exist asymmetrically distorted types for introducing some important binocular visual properties (e.g., binocular rivalry, visual discomfort) comparing with 2D IQA databases. It makes the research more difficult and interesting.

The rest of this Chapter is organized as follows: Sect. 3.2 will discuss the unique visual properties: binocular vision. Section 3.3 will formally introduce stereoscopic image quality assessment, involving the principle of SIQA and existing 3D databases. Detailed discussions about the general designed frameworks of current objective SIQA methods are provided in Sect. 3.4. And summary and conclusions are drawn in Sect. 3.5.

3.2 Binocular Vision

Compared with 2D image process, binocular vision is one of the most important visual characteristics for stereoscopic images. In particular, binocular vision is a kind of unique visual characteristic for stereo images. When attempting to integrate two monocular views into stereopsis, binocular visual properties will occur in human's brain. However, it is also a complex and puzzled image process involving the fields of psychology and physiology. Seeking a proper model for simulating the procedure of binocular vision is difficult. Thus, even though advancing progress has been made in 2D IQA, it is still an open and challenging area for quality evaluation of stereo images. Fortunately, binocular vision process has aroused keen interests of scientists in various fields that leading to many experiments and conclusions which allows a better understanding of the stereo property (Hubel and Wiesel 1962; Krüger et al. 2013). Thus, it is necessary to fully understand binocular vision before starting the research of SIQA. There are numerous discoveries of binocular vision that have been proved in physiological experiments, where we focus on the binocular visual behaviors which describe the visual inputs integration process. In subsequent sections, we will introduce binocular vision in four aspects: binocular disparity, binocular fusion, binocular rivalry, ocular dominance and visual discomfort.

3.2.1 Binocular Disparity

As described at the beginning of this chapter, stereopairs captured by two eyes are slightly different horizontally because that human eyes are separated by a distance of approximately 6.3 cm on average (Schreer et al. 2005), which is called binocular disparity. The different contents between two views provide cues about the relative distances of viewed objects (Backus et al. 1999). Obviously, the effectiveness of binocular disparity is quite strong at close distances. As the distance between objects and viewer increases, the location difference of the same objects in two views becomes smaller and vice versa. Figure 3.3 gives an intuitive illustration about this fact, from which we can easily find that the value of dis_1 is greater than dis_2, because the object B is farther from the viewer than A.

Hence, given the left and right views of a stereoscopic image, human brain can easily merge them into a corresponding cyclopean image and further distinguish which objects in the stereopair are closer to us and which are farther away. Generally, one image of stereopairs can be restored according to the disparity information and another monocular image of the stereopairs. In another words, a corresponding cyclopean image can be generated from the two monocular views. When both monocular views are corrupted by different distortion types or levels, the disparity information will also be distorted. So that the brain can hardly attain proper depth perception for recognizing the actual distance of objects. This will ultimately be reflected in estimation of subjective quality scores. Consequently, it is reasonable to think that

3.2 Binocular Vision

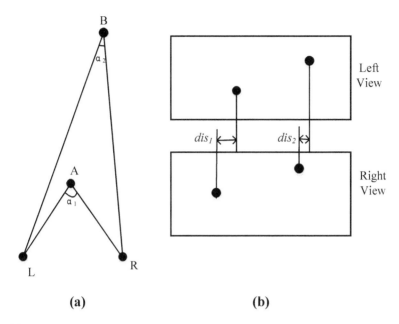

Fig. 3.3 Difference of object depths reflected in **a** different viewing angles of left and right eyes and **b** different position shifts in left and right views

the disparity between the left and right eye images has an important impact on visual quality assessment.

3.2.2 Binocular Fusion

Due to the average 6.3 mm distance between the left and right eyes in human beings, there are slightly different object scene in the two views captured by the two eyes (Schreer et al. 2005). Fortunately, human brain merges the two views into a corresponding intermediate image with binocular depth perception (also known as cyclopean image) instead of treating them separately, which is called as binocular fusion. In reality, humans always perceive visual information from the intermediate image instead of two separated monocular views, and even humans do not realize it, which is owing to that binocular fusion exists every time except for closing one eye (Julesz 1971). At the meanwhile, due to the overlapping areas of the two monocular views, humans can easily sense the depth information of each object in the scene at all times (Fahle 1987). Thus, the modeling of binocular fusion is fundamental for SIQA research.

So far, the mechanism of binocular fusion has been studied by researchers for decades, and some interesting binocular visual behaviors have been discovered in some literatures. The famous binocular visual behaviors include, but are not limited to

Fechner's paradox and cyclopean perception. From Fechner's paradox, the perceived visual information of one eye may be reduced when the other eye receiving a week stimulus. Why this binocular behavior is called paradox is that the visual information ultimately perceived by humans is not positively linear with the input visual information. Cyclopean perception describes the fact that the visual information and scenes perceived by two eyes are similar with that only perceived by one eye when we close the other eye, which is a very common visual phenomenon in our daily life. In addition, (Ding et al. 2009) found a binocular behavior existed in persons with amblyopia, which is called interocular contrast gain enhancement. The authors discovered that the slight stimulus perceived by the non-dominant eye (NDE) does not inhibit the visual stimulus perceived by the dominant eye (DE). In contrast, the NDE could enhance the DE, making it apparent under certain conditions (e.g., persons with amblyopia). Because of the existence of these discussed interesting binocular behaviors, the research of binocular fusion becomes not so simple. When constructing the mathematical model of binocular fusion, the core is considering whether it can explain Fechner's paradox and cyclopean perception well.

3.2.3 Binocular Rivalry

When humans perceive two monocular views with different quality or nature scenes, humans' brains cannot directly merge them into a perfect cyclopean image with correct depth information. More generally, the two views captured by the left and right eyes will compete with each other, and eventually the visual perception in human's brain will be dominated by the "dominant" one of the two views, which is called binocular rivalry. As early as in 1960, an informal conclusion was obtained that the perceived quality of a stereoscopic image seems to be determined by the view with relatively higher image quality (Julesz 1960). Therefore, the "dominant" image was once considered the image with relatively higher quality of the two. However, an opposite conclusion in later research was concluded from the observation that the perceived quality of a stereopair is mainly dominated by the lower-quality monocular view when both of them occurring the distortion of JPEG compression (Meegan et al. 2001), illustrated in Fig. 3.4b. In contrast, for the distortion of blurring, the perceived stereoscopic image quality tends to be close to the higher-quality view, which is shown in Fig. 3.4a. The definition of "dominant" image seems to be more complicated than before. Todays, the researchers assume that human visual system (HVS) is more likely interested in the view with more visual information rather than that with higher quality. Therefore, the definition of binocular rivalry can be re-written as that the generation of a cyclopean image is eventually dominated by the view with more information, in other words, with larger energy, which also meets the visual attention mechanism of HVS.

The concept of binocular rivalry is very important in measuring the perceived quality of a distorted stereopair. If binocular rivalry is not existed in HVS, the quality of a stereoscopic can be simply recognized as the average of quality scores of the two

3.2 Binocular Vision

Fig. 3.4 The perceived quality of a stereoscopic image is dominated by the left view (higher quality) in (**a**) and dominated by the right view (lower quality) in (**b**)

monocular views. Binocular rivalry, as one of the most unique visual mechanisms in stereo vision, makes the SIQA tasks more complex and challenging. And even when creating the LIVE 3D Database Phase II, the researchers introduced a special distortion type, named asymmetrical distortion, whose main purpose was to consider the influence of binocular rivalry. In the following chapters, we will explore the effects of binocular rivalry furtherly and discuss how to properly consider binocular rivalry in SIQA tasks.

3.2.4 Ocular Dominance

When we introduced the binocular phenomenon of interocular contrast gain enhancement in Sect. 3.2.2, we mentioned the DE and the NDE, but did not give a clear definition of them. In this section, we will explore them in detail.

Ocular dominance, or eye dominance, refers to the HVS preference to the visual input of one eye compared to that of the other eye (Khan and Crawford 2001). We can understand ocular dominance according to the utilization of "handedness". The previous studies had explored the fact that about two-thirds of human beings in the world are born with right-eye dominance, and the remaining about one-third are with left eye dominance (Ehrenstein et al. 2005). In addition, there are little people that belongs to neither of the two classes.

The definition of handedness can help to understand the concept of ocular dominance in some way. In the world, there are quite a few people with a more developed left cerebral hemisphere to control their right-hand preferences. They will prefer to use the right hand to solve various problems in their daily life and work. Similarly, there are still some people using right cerebral hemisphere to control a left-hand preference. Motivated by the definition of handedness, ocular dominance also can be recognized as "eyedness". However, the concept of eyedness does not always match the definition of handedness (Porac and Coren 1975). In ocular dominance, both left and right cerebral hemispheres control the dominant eye and the non-dominant eye, and each hemisphere controls half of both retinas, making both eyes work together in capturing a specific object. The dominant eye locates a particular object laterally (i.e., from left to right or from up to down), while the non-dominant eye is responsible for locating the same object longitudinally (i.e., near or far). The two eyes work together to locate the focus of nature scenes bicamerally and obtain 3D visual information. Moreover, the dominance of the dominant eye over the non-dominant eye may not be changeless. For some people, the relative dominance between the left and right eyes may change when the gaze direction moves or changes.

The existence of ocular dominance increases the difficulty of SIQA research. So far, there is not a clear recognition whether the left-eye dominance or the right-eye dominance is adopted in SIQA for better helping the performance improvement, although the fact has been found that people with the right-eye dominance is the majority in the world. However, most existing SIQA methods neglected the influence of ocular dominance, but do not achieve quite poor prediction performance. This may be attributed to two aspects. On the one hand, the collection of subjective quality scores is a mathematical statistical process (e.g., MOS and DMOS), which eliminates the errors caused by individual differences to the greatest extent, including ocular dominance. On the other hand, the influence of ocular dominance can be neglected when we take binocular rivalry into account, because the influence of binocular rivalry is greater than that of ocular dominance, or ocular dominance is contained in binocular rivalry more or less. Therefore, what we can do now is only to ignore the effect of ocular dominance in guaranteeing the prediction performance. But when we encounter a bottleneck of SIQA research in the future, we may only consider the effect of ocular dominance to further improve the performance of SIQA algorithms at this time.

3.2.5 Visual Discomfort

The last binocular visual property we would like to introduce is visual discomfort. Visual discomfort will occur with the symptoms (e.g., headache, asthenopia and fatigue) when humans force their eyes to focus on a fixed focal distance for a long time, or the object distance perceived by humans is seriously distorted (Lambooij et al. 2009; Choi et al. 2012). In addition, there are many reasons that cause visual discomfort, which is studied by many researchers for decades. (Kooi and Toet 2004)

3.2 Binocular Vision

found that flawed presentations of horizontal difference between the left and right views captured by human eyes, such as excessively large or otherwise unnatural disparities, could lead to severe visual discomfort. Tyler et al. (2012) explored that visual discomfort could be caused by misalignment of stereopairs with regard to vertical and torsional disparities. Generally, statistical features related to depth information are more likely to cause severe visual discomfort, such as larger depth, wider depth range, nonuniform depth distribution, etc. In other words, the degree of visual discomfort mainly depends on the distortion of disparity information of viewed stereopairs in the absence of geometrical distortions and window violations (Yano et al. 2002).

The relationship between visual discomfort and SIQA research is not clear in the current time. But it is ensured that the former will affect the latter, although some researchers tried their best to take the visual experiences of the observers into account when they created SIQA databases (Chen et al. 2013b). A reasonable evidence for this conclusion is that distortions in stereoscopic images can also cause visual discomfort except the observer's poor viewing experience. Due to stereo-pair's distortion, humans may not be able to estimate correct depth information and obtain a stereopsis, which definitely affects the degree of visual discomfort. To conclude, visual discomfort is one of the most important binocular properties closely related to the quality of stereopairs, which demands to be considered and studied in SIQA research. In Chap. 5, we will introduce visual discomfort in detail and its application in SIQA research.

3.3 Subjective Stereoscopic Image Quality Assessment

3.3.1 Principle

Now let us introduce the SIQA more formally. Like 2D IQA as explained in Chap. 2, SIQA also can be addressed by two ways: subjective system and objective system. Accurate subjective quality scores can be generated by subjective quality assessment of several subjects in the controlled laboratory. However, it is time-consuming and laborious. Therefore, objective SIQA are develop urgently to explore 3D image quality in real time. Following the principle of whether to rely on reference stereopair information, SIQA also can be divided into three categories: FR, RR and NR SIQA, in which NR SIQA is more promising and interesting in practical applications. Another common point between 2D IQA and 3D IQA is to construct objective IQA models to pursue the consistency between the designed objective algorithm and subjective perception by extracting visual features that can reflect the perceptual characteristics of HVS.

After introducing what 2D IQA and 3D IQA have in common, there are differences between them needed to be introduced, in which we only discuss about one important difference: the introduction of asymmetrical distortion in 3D IQA. As described in the previous section, a stereopair consists of two monocular views: the left and right images with slightly difference in horizontal axis. When the two views are distorted asymmetrically (i.e., the left and right views are altered by the same distortion type with different distortion levels, or even the different distortion types with various levels), the property of binocular rivalry will occur in assessing the quality of stereopairs, which makes SIQA research more meaningful and difficult. That's the biggest difference between 2D IQA and 3D IQA, which leads us to focus our research on binocular vision including binocular combination, disparity difference, binocular rivalry and binocular discomfort. For example, when evaluating the quality of 3D images, it is intuitive to think that we can use 2D IQA algorithm to calculate the quality of left and right views respectively and take its average value as the final quality of 3D images (Yang et al. 2010). However, the subsequent studies demonstrate the imprecision of this SIQA method, especially for asymmetric distortion. Therefore, assessing image quality of stereopairs is not simply a mathematical integration of existing 2D IQA models, but requires more in-depth research for exploring binocular properties in stereopsis, which is very challenging.

3.3.2 Databases

To the best of our knowledge, there are 10 publicly quality databases of stereoscopic images that have been proposed for validation of objective SIQA methods. Table 3.1 lists the basic information of these publicly available databases. In the following sections, some of them will also be introduced and discussed with detailed descriptions.

(1) IRCCyN/IVC 3D Images Database (Benoit et al. 2008a)
   The IRCCyN/IVC 3D images database can be recognized as the earliest subject-rated image database of the SIQA research community with our best knowledge, which were built from the team Image and Video-Communications (IVC), Institut de Recherche en Communication et Cybernétique de Nantes (IRCCyN), Nantes, France. The database contains 90 distorted stereoscopic images created from 6 reference versions symmetrically. 15 corresponding distorted images for each reference stereopair are generated by three distortion types with various distortion levels including JPEG 2000 compression (JP2K), JPEG compression, and Gaussian blur (GB). Like LIVE database in 2D IQA research, each distorted stereoscopic image is also given a difference mean opinion score (DMOS) value as the subjective quality score through the subjective test of 17 subjects, where larger DMOS value means better image quality.
(2) MICT 3D Image Quality Evaluation Database (Sazzad et al. 2009)
   The builders are from Graduate School of Science and Engineering, University of Toyama, Japan. The database contains 490 symmetrically and asymmetrically distorted images created from 10 pristine stereopairs, in which JPEG compression is the only considered distortion type. Mean opinion scores (MOSs) are

3.3 Subjective Stereoscopic Image Quality Assessment

Table 3.1 Basic information about the databases

	Year	Nationality	Reference images	Image sizes	Distorted images	Symmetrical	Asymmetrical	Distortion types	Subjects	Subjective quality
IRCCyN/IVC	2008	France	6	512 × 448	90	Yes	No	3	17	DMOS
MICT	2009	Japan	10	640 × 480	490	Yes	Yes	1	24	MOS
NBU-I	2009	China	10	1252 × 1110 to 1390 × 1110	400	No	Yes	4	20	DMOS
NBU-II	2011	China	12	480 × 270 to 1024 × 768	312	Yes	No	5	26	DMOS
TJU	2009	China	30	320 × 240 to 1024 × 768	270	Yes	No	3	N/A	DMOS
MMSPG	2010	Switzerland	N/A	1920 × 1080	100	Yes	No	1	20	MOS
IEEE-SA	2012	N/A	13	N/A	650	Yes	No	5	N/A	DMOS
LIVE-I	2013	USA	20	640 × 360	365	Yes	No	5	32	DMOS
LIVE-II	2013	USA	8	640 × 360	360	Yes	Yes	5	33	DMOS
MCL-3D	2014	USA	9	1024 × 728 to 1920 × 1080	684	Yes	No	6	270	MOS
Waterloo-I	2015	Canada	6	1920 × 1080	330	Yes	Yes	3	24	MOS
Waterloo-II	2015	Canada	10	1920 × 1080	460	Yes	Yes	3	22	MOS

N/A means the unavailable information which is not introduced in the original paper computed for each distorted stereo image after the subjective test of 24 non-expert subjects as the subjective evaluation results. Higher MOS values represent higher visual quality.

(3) NBU 3D Image Quality Assessment Database Phase I and II (Wang et al. 2009; Zhou et al. 2011)

NBU 3D database phase I was created from 10 pristine stereopairs in 2009. The database contains 410 asymmetrically distorted stereoscopic images altered by JP2K, JPEG, white noise (WN), and GB, where asymmetrical distortion means that the right view of stereopair is distorted while the left view is kept undistorted. Compared with phase I, phase II consists of 312 symmetrically distorted images generated from 12 reference versions with five types of image distortion, in which H.264 compression is added for a new distortion type.

(4) TJU 3D Image Quality Assessment Database (Yang et al. 2009)

This database also contains only symmetrically distorted images, created by the stereo image database of school of electronic and information engineering, Tianjin University (TJU), China. 270 images with the distortion types of JP2K, JPEG, and WN are included in the database, where DMOS as the subjective score is assigned for each stereopair.

(5) MMSPG 3D Image Quality Assessment Database (Goldmann et al. 2010)

The database is created from Multimedia Signal Processing Group (MMSPG), Ecole Polytechnique Fédérale de Lausanne, Switzerland. This database is quite unique that different camera distances between left and right views (i.e., from 10 to 60 cm) are set as the distortion type of the images.

(6) IEEE-SA 3D Image Quality Assessment Database (IEEE 2012)

650 distorted S3D image pairs are generated from thirteen reference stereopairs with five distortion types (i.e., 130 each for JPEG, JP2K, WN, BLUR, and FF). The corresponding disparity maps and DMOS values for each distorted stereopair are provided in the database.

(7) LIVE 3D Image Quality Assessment Database Phase I and II (Moorthy et al. 2013; Chen et al. 2013b)

Both LIVE 3D database phase I and phase II are constructed by the famous Laboratory for Image and Video Engineering (LIVE), University of Texas at Austin, United States, which currently have become the most common database for evaluating the performance of objective SIQA methods. 360 and 365 distorted images are contained in phase I and phase II respectively, which are distorted by five distortion types: JPEG, JP2K, white noise (WN), GB, and fast-fading in Rayleigh Channel (FF), same as in LIVE 2D IQA database. Different from phase I, phase II introduces the asymmetrical distortion type for each reference stereopair to simulate binocular rivalry, which is extremely challenging. Similar, the two databases are allocated DMOS values for each distorted stereopair as subjective quality scores.

(8) MCL 3D Image Quality Database Phase (Song et al. 2015)

The database contains 684 symmetrically distorted stereopairs created from the University of Southern California, USA. Nine image-plus-depth sources were first selected, and a depth-image-based rendering technique was used to render

3D images. Six distortion types with four levels were applied to either the texture 3D image or depth image prior to 3D image rendering, including GB, WN, down sampling blur (SBLUR), JPEG, JP2K, and transmission error (TERROR).

(9) Waterloo-IVC 3D Image Quality Database Phase I and II (Wang et al. 2015)
Both the two databases are built by researchers with the Department of Electrical and Computer Engineering, University of Waterloo, Waterloo, Canada. Each stereo image is corrupted by three types of distortion symmetrically or asymmetrically, including WN, GB, and JPEG. In addition, a special distortion type, called the asymmetrical hybrid distortion, is designed in the database, where the left and right views of stereopair are altered by different distortion types with different distortion levels. Finally, 330 and 460 symmetrically and asymmetrically distorted stereopairs are generated in phase I and phase II, respectively. Moreover, the two databases also provide several distorted single-view images for exploring the relationship between the perceptual quality of single-views and corresponding stereoscopic images.

In addition, to better benchmark the performance of objective SIQA methods, three general criteria used in 2D IQA are also adopted in 3D IQA, including Pearson's Linear Correlation Coefficient (PLCC), Spearman's Rank-order Correlation Coefficient (SRCC) and Root Mean Squared Error (RMSE). PLCC and RMSE are usually estimated for representing the prediction accuracy and consistency, while SRCC can be recognized as a criterion of prediction monotonicity. Higher values of PLCC and SRCC, and lower value of RMSE indicate better performance. For more information of the three general criteria, please refer to Chap. 2.

3.4 General Frameworks of SIQA Models

There are four kinds of approaches that commonly adopted in SIQA research community to exploring the quality of stereo images over the past few years. Here we will introduce these general frameworks briefly according to the research history, and more detailed introduction and analysis will be given in the following Chapters.

The first insight is directly to extend classical 2D IQA models to each of views, respectively. Then the obtained two monocular global scores are integrated into the final quality rating (Benoit et al. 2008b; Yasakethu et al. 2008). In later research, considering stereo quality factors, like depth perception, the framework can be further improved by adding statistical characteristics of disparity map or other binocular visual factors (You et al. 2010; Akhter et al. 2010) to increase the prediction performance, as depicted in Fig. 3.5.

Afterwards, to explore the interrelation of the two views of stereopair in human brain, many researchers began to focus on the theory that a cyclopean image will be generated from the two views through the binocular behaviors of simple and complex cells before it is perceived by human brain (Zhao et al. 2016; Zhou et al. 2017a, b; Ding and Zhao 2017; Sun et al. 2018). Typically, (Chen et al. 2013b) developed a

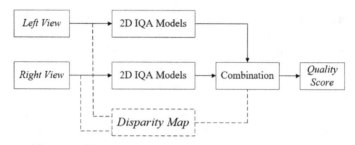

Fig. 3.5 A simplified illustration of what an element in a disparity map represents

binocular fusion model in which local energy response of left and disparity-shift right views were calculated as corresponding weighting coefficients to generate the corresponding cyclopean image. Then, the generated cyclopean image could apply some well-known 2D IQA models to obtain the final quality score. Note that, the procedure of cyclopean image construction contains several stereo visual properties, including depth perception, binocular fusion and binocular rivalry behaviors. Figure 3.6 shows the SIQA framework with cyclopean image construction.

There are many human visual properties in stereopairs perceived by human eyes except for the unique binocular properties. In 2D IQA, the behaviors of HVS have been thoroughly introduced and analyzed for improving prediction performance, including visual saliency, just noticeable difference (JND) (Moorthy and Bovik 2009; Zhang and Li 2012; Toprak and Yalman 2017). Therefore, it has become an important research target of 3D IQA to combine binocular mechanisms and human visual properties to analyze stereo images, of which the framework is utilized in Fig. 3.7. Assisted by human visual properties, significant progresses have been achieved in 3D IQA research community (Shao et al. 2016a, b; Xu et al. 2017; Gu et al. 2019), demonstrating the importance and necessity of HVS.

The last type of approaches we would like to introduce is deep learning-based SIQA. With the development of deep learning in many image processing fields (e.g., image recognition, image classification and object detection), many researchers began to attempt apply shallow convolutional neural networks (CNNs) into IQA fields to learn visual representations relating to image quality instead of hand-crafted visual features (Lv et al. 2016; Yang et al. 2019). Compared to traditional image processing

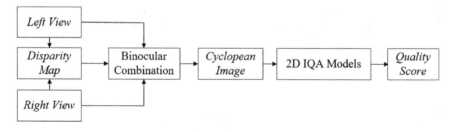

Fig. 3.6 SIQA framework with cyclopean image construction

3.4 General Frameworks of SIQA Models

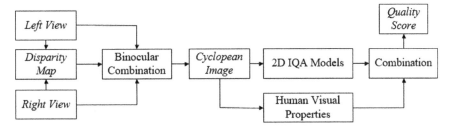

Fig. 3.7 3D framework combining binocular mechanisms and human visual properties

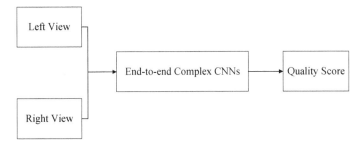

Fig. 3.8 End-to-end CNN-based SIQA model

approaches, CNNs can automatically learn task-related visual features by optimizing networks' parameters in larger image datasets during training. Figure 3.8 shows the procedure of CNN-based SIQA models optimized in an end-to-end manner. After alleviating the obstacle of insufficient training data, various deeper CNN models are designed and adopted to explore the quality of 2D and 3D images, which delivers excellent predicted results (Oh et al. 2017; Ding et al. 2018; Zhou et al. 2019; Sun et al. 2020).

3.5 Summary

In the past few years, significant works have been made in 2D IQA research community. With the urgent increase of the demand for 3D image quality, the research of SIQA has been conducted recently and achieves considerable results. Compared with plant images, richer 3D visual information of stereopairs need to be explored and analyzed before employing 2D IQA thoughts. Fortunately, there are many studies in SIQA tasks which take binocular vision into consideration. In this chapter, four kinds of SIQA approaches are introduced briefly according to its development history, which can give readers a better understanding in SIQA research community. In addition to directly employ 2D IQA models into each of monocular view, more science ways have recognized the fact that a cyclopean image can be generated from two monocular views by human's brain before obtaining visual information. The insight to construct cyclopean image for evaluating stereo image quality starts the research interest of 3D IQA. By jointly considering both binocular mechanisms and human visual properties, SIQA could achieve a more promising result than before. Finally, due to its powerful learning ability, CNNs are adopted in the fields of SIQA to expect obtaining a good prediction result.

As discussed above, research on SIQA is getting deeper and better. However, there's still a lot of confusion about binocular vision needed to be explored deeply. For example, binocular property is not only simulated by the simple binocular fusion models proposed in (Chen et al. 2013b). Therefore, more detailed research on stereo images has yet to be done. From the previous reviews, it is believed that significant progress and breakthroughs will be made for objective SIQA research in the near future.

References

Akhter R, Sazzad ZMP, Horita Y, Baltes J (2010) No-reference stereoscopic image quality assessment. Proc SPIE 7524:75240T

Backus B, Banks M, van Ee R, Crowell J (1999) Horizontal and vertical disparity, eye position, and stereoscopic slant perception. Vision Res 39(6):1143–1170

Benoit A, Le Callet P, Campisi P, Cousseau R (2008a) Quality assessment of stereoscopic images. EURASIP J Image Video Process: 659024

Benoit A, Le Callet P, Campisi P, Cousseau R (2008b) Using disparity for quality assessment of stereoscopic images. In: IEEE international conference on image processing, San Diego, United States, pp 389–392

Chen M-J, Cormack LK, Bovik AC (2013a) No-reference quality assessment of natural stereopairs. IEEE Trans Image Process 22(9):3379–3397

Chen M-J, Su C-C, Kwon D-K, Cormack LK, Bovik AC (2013b) Full-reference quality assessment of stereopairs accounting for rivalry. Sig Process Image Commun 28:1143–1155

Choi J, Kim D, Choi S, Sohn K (2012) Visual fatigue modeling and analysis for stereoscopic video. Opt Eng 51(1):017206

Ding J, Klein S, Levi D (2009) Binocular combination in amblyopic vision. J Vis 9(8):274

Ding Y, Zhao Y (2017) No-reference quality assessment for stereoscopic images considering visual discomfort and binocular rivalry. Electron Lett 53(25):1646–1647

Ding Y, Deng R, Xie X, Xu X, Chen X et al (2018) No-reference stereoscopic image quality assessment using convolutional neural network for adaptive feature extraction. IEEE Access 6:37595–37603

Ehrenstein WH, Arnold-Schulz-Gahmen BE, Jaschinski W (2005) Eye preference within the context of binocular function. Graefe's Arch Clin Exp Ophthalmol 243(9):926–932

Fahle M (1987) Two eyes, for what? Naturwissenchaften 74(8):383–385

Goldmann L, De Simone F, Ebrahimi T (2010) Impact of acquisition distortions on the quality of stereoscopic images. In: International workshop on video processing, quality metrics, and consumer electronics, Scottsdale, Arizona, USA, 13–15

Gu Z, Ding Y, Deng R, Chen X, Krylov AS (2019) Multiple just-noticeable-difference based no-reference stereoscopic image quality assessment. Appl Opt 58(2):340–352

Hubel DH, Wiesel TN (1962) Receptive fields, binocular interaction and functional architecture in cats visual cortex. J Physiol 160(1):106–154

References

IEEE (2012) IEEE-SA stereo image database. [Online]. Available: http://grouper.ieee.org/groups/3dhf/

Julesz B (1960) Binocular depth perception of computer-generated patterns. Bell Syst Tech J 39(5):1125–1162

Julesz B (1971) Foundations of cyclopean perception. The University of Chicago Press, Chicago

Khan AZ, Crawford JD (2001) Ocular dominance reverses as a function of horizontal gaze angle. Vis Res 41(14):1743–1748

Kooi FL, Toet A (2004) Visual comfort of binocular and 3D displays. Displays 25(2–3):99–108

Krüger N, Janssen P, Kalkan S, Lappe M, Leonardis A et al (2013) Deep hierarchies in the primate visual cortex: what can we learn for computer vision? IEEE Trans Pattern Anal Mach Intell 35(8):1847–1871

Lambooij MTM, Ijsselsteijn WA, Heynderickx I (2009) Visual discomfort in stereoscopic displays: a review. J Imaging Sci Technol 53(3)

Lv Y, Yu M, Jiang G, Shao F, Peng Z et al (2016) No-reference stereoscopic image quality assessment using binocular self-similarity and deep neural network. Sig Process Image Commun 47:349–357

Meegan DV, Stelmach LB, Tam WJ (2001) Unequal weighting of monocular inputs in binocular combination: implications for the compression of stereoscopic imagery. J Exp Psychol: Appl 7(2):143–153

Moorthy AK, Bovik AC (2009) Visual importance pooling for image quality assessment. IEEE J Sel Topics Signal Process 3(2):193–201

Moorthy AK, Su C-C, Mittal A, Bovik AC (2013) Subjective evaluation of stereoscopic image quality. Sig Process Image Commun 28(8):870–883

Oh H, Ahn S, Kim J, Lee S (2017) Blind deep S3D image quality evaluation via local to global feature aggregation. IEEE Trans Image Process 26(10):4923–4936

Porac C, Coren S (1975) Is eye dominance a part of generalized laterality? Percept Mot Skills 40(3):763–769

Sazzad ZMP, Yamanaka S, Kawayokeita Y, Horita Y (2009) Stereoscopic image quality prediction. In: International workshop on quality of multimedia experience, San Diego, United States, pp 180–185

Schreer O, Kauff P, Sikora T (2005) 3D video communication: algorithms concepts and real-time systems in human centred communication, Wiley

Shao F, Li K, Lin W, Jiang G, Yu M, Dai Q (2015) Full-reference quality assessment of stereoscopic images by learning binocular receptive field properties. IEEE Trans Image 24(10): 2971–2983. https://doi.org/10.1109/TIP.2015.2436332

Shao F, Lin W, Jiang G, Dai Q (2016a) Models of monocular and binocular visual perception in quality assessment of stereoscopic images. IEEE Trans Comput Imaging 2(2):123–135

Shao F, Tian W, Lin W, Jiang G, Dai Q (2016b) Toward a blind deep quality evaluator for stereoscopic images based on monocular and binocular interactions. IEEE Trans Image Process 25(5):2059–2074

Song R, Ko H, Kuo CCJ (2015) MCL-3D: A database for stereoscopic image quality assessment using 2D-image-plus-depth source. J Inf Sci Eng 31(5):1593–1611

Sun G, Ding G, Deng R, Zhao Y, Chen X et al (2018) Stereoscopic image quality assessment by considering binocular visual mechanisms. IEEE Access 6:511337–511347

Sun G, Shi B, Chen X, Krylov AS, Ding Y (2020) Learning local quality-aware structures of salient regions for stereoscopic images via deep neural networks. IEEE Trans Multim 2020:1

Toprak S, Yalman Y (2017) A new full-reference image quality metric based on just noticeable difference. Comput Stand Interfaces 50:18–25

Tyler CW, Likova LT, Atanassov K, Ramachandra V, Goma S (2012) 3D discomfort from vertical and torsional disparities in natural images. Proc SPIE-Soc Photo-Opt Instrum Eng 8291:1–9

Wang X, Yu M, Yang Y, Jiang G (2009) Research on subjective stereoscopic image quality assessment. Proc SPIE 7255:725509

Wang J, Rehman A, Zeng K, Wang S, Wang Z (2015) Quality prediction of asymmetrically distorted stereoscopic 3D images. IEEE Trans Image Process 24(11):3400–3414

Xu X, Zhao Y, Ding Y (2017) No-reference stereoscopic image quality assessment based on saliency-guided binocular feature consolidation. Electron Lett 53(22):1468–1470

Yang J, Hou C, Xu R (2010) New metric for stereo image quality assessment based on HVS. Int J Image Syst Technol 20(4):2010

Yang J, Hou C, Zhou Y, Zhang Z, Guo J (2009) Objective quality assessment method of stereo images. In: 2009 3DTV-Conference: the true vision – capture, transmission and display of 3D video, Potsdam, Germany

Yang J, Zhao Y, Zhu Y, Xu H, Lu W et al (2019) Blind assessment for stereo images considering binocular characteristics and deep perception map based on deep belief network. Inf Sci 474:1–17

Yano S, Ide S, Mitsuhashi T, Thwaites H (2002) A study of visual fatigue and visual discomfort for 3D HDTV/HDTV images. Displays 23(4):191–201

Yasakethu SLP, Hewage CTER, Fernando WAC et al (2008) Quality analysis for 3D video using 2D video quality models. IEEE Trans Consum Electron 54(4):1969–1976

You J, Xing L, Perkis A, Wang X (2010) Considering binocular spatial sensitivity in stereoscopic image quality assessment. In: Proceedings of the international workshop of video processing, Firenze, Italy, pp 61–66

Zhang L, Li H (2012) SR-SIM: a fast and high performance IQA index based on spectral residual. In: Proceedings of IEEE international conference on image processing, Lake Buena Vista, Florida, pp 1473–1476

Zhao Y, Ding Y, Zhao X (2016) Image quality assessment based on complementary local feature extraction and quantification. Electron Lett 52(22):1849–1851

Zhou J, Jiang G, Mao X, Yu M, Shao F et al (2011). Subjective quality analyses of stereoscopic images in 3DTV system. IEEE international conference on visual communications and image processing, Tainan, Taiwan, pp 1–4

Zhou W, Chen Z, Li W (2019) Dual-stream interactive networks for no-reference stereoscopic image quality assessment. IEEE Trans Image Process 28(8):3946–3958

Zhou W, Qiu W, Wu M (2017a) Utilizing dictionary learning and machine learning for blind quality assessment of 3-D images. IEEE Trans Broadcast 63(1):404–415

Zhou W, Zhang S, Pan T, Yu L, Qiu W et al (2017b) Blind 3D image quality assessment based on self-similarity of binocular features. Neurocomputing 224:128–134

Chapter 4
SIQA Based on 2D IQA Weighting Strategy

Abstract As image quality assessment (IQA) methods for plant images have been explored thoroughly, some SIQA algorithms apply 2D IQA methods on both stereoscopic views independently and then combine the two scores to obtain an overall quality score by a dedicated strategy. The early algorithms only combine the two scores simply to obtain the final quality score, and some improved algorithms utilize both the stereoscopic views and the depth/disparity information. All of these algorithms could achieve fairly good performance. The mainstream and state-of-art SIQA based 2D IQA weighting strategy are introduced in detail in this chapter.

Keywords Stereoscopic image quality assessment · Disparity information · Quality score

4.1 Introduction

In Chap. 2, image quality assessment (IQA) methods for plant images have been explored and discussed, where the objects for quality evaluation were individual digital images. And some challenges from 2D to 3D images were also analyzed in the previous chapter. In the early studies, stereoscopic image quality assessment (SIQA) methods were relatively less explored by researchers, partially due to the difficulty in viewing commercially available 3D contents, as well as the problems of surrounding visual discomfort and fatigue after long-term stereoscopic viewing (Lambooij et al. 2009). However, with the development of technology for 3D image/video display and communication, 3D stereo media has greatly increased the popularity and availability for human consumption, and applications can range from entertainment (e.g., 3D cinema, 3D game, and 3D television) to many specialized purposes (e.g., 3D robot navigation, 3D remote education, and 3D medical imagery). Therefore, there is an urgent demand for spreading IQA algorithms into 3D (stereo) images.

As mentioned in the last chapter, a stereoscopic image contains two slightly different views, each of which is projected onto the human retina. Human visual system (HVS) does not take the combination of monocular visual signals as simple summation, but goes through a very complex process of binocular fusion and rivalry between two views to generate a merged cyclopean view. The quality of 3D image is not only affected by the degradation level of each individual left and right view, but also by the binocular visual experience (Braddick 1979). The earliest algorithm was to simply apply 2D IQA methods on both stereoscopic views independently and then combine the two scores to obtain an overall quality score (Yasakethu et al. 2008; Hewage et al. 2008; Gorley and Holliman 2008). Then many researchers proposed some improved algorithms that the main insight was to utilize both the stereoscopic views and the depth/disparity information (Benoit et al. 2008; Kaptein et al. 2008; Goldmann and Ebrahimi 2010). All of these algorithms could achieve fairly good performances only on predicting quality of symmetrically distorted stereoscopic images. As for the two stereoscopic views include different amounts and types of distortion (also called the asymmetrically distorted stimuli), some researchers have proposed approaches to model the 3D quality assessment behavior of the HVS based on binocular fusion and suppression properties (Qi et al. 2015; Ahmed and Larabi 2014). The mainstream and state-of-art SIQA based 2D IQA weighting algorithms are introduced in this chapter.

4.2 SIQA Algorithms Based on 2D IQA Methods

The earliest work on SIQA was based on 2D image quality objective metrics. In short, quality scores on both images (right and left) of the stereo content were evaluated by means of the typical 2D image quality metrics, such as SSIM, UQI, C4, and RRIQ (Wang et al. 2004; Wang and Bovik 2002; Carnec et al. 2003; Wang and Simoncelli 2005). Then the obtained two scores for both of the left and right views could be combined to a single metric for the quality assessment of the stereo image. There are three different combination approaches that have been proposed (Boev et al. 2009): "average", "main eye" and "visual acuity" approaches.

The simplest combination method is the "average" method, which directly averaged the quality scores of left and right views to obtain the final scores. The objective scores Q_O could be expressed as follows (Campisi et al. 2007):

$$Q_O = \frac{1}{2}(Q_L + Q_R) \quad (4.1)$$

where Q_L and Q_R are objective scores for the left and the right images respectively.

The "main eye" method considered the main eye of each observer and put it into the formula:

4.2 SIQA Algorithms Based on 2D IQA Methods

$$Q_O = \frac{1}{N_L + N_R}(N_L Q_L + N_R Q_R) \tag{4.2}$$

where N_L and N_R denote the number of observers whose main eye is the left or the right one, respectively. However, the quantitative evaluation of this method showed that no significant performance improvement had been achieved compared with "average" method.

The "visual acuity" method made use of the visual acuity of the observers to weight the left and the right image. Specifically, the objective score could be obtained by

$$Q_O = \frac{1}{N_L + N_R}\left(\frac{\sum_i A_{L,i}}{A_{\max}}Q_L + \frac{\sum_i A_{R,i}}{A_{\max}}Q_R\right) \tag{4.3}$$

where $A_{L,i}$ and $A_{R,i}$ represent the visual acuity of the left and right eye of the i-th observer respectively, and $A_{\max}$ represents the maximum value for the visual acuity that has been set equal to ten. The weights for Q_L and Q_R had been observed to be equal to 0.43 and 0.57, which were not significantly different from "average" method. Therefore, the results of experiments showed that this metric had led to no performance improvement compared with the "average" method.

4.3 SIQA Algorithms Employ the Disparity Information

As described in Chap. 3, human eyes are horizontally separated by a slight distance, resulting that the image contents perceived by two eyes are different in horizontal axis. Researchers have been aware of the importance of the difference information, also called disparity information. In order to improve the accuracy of stereo image quality scores, some research took a different perspective by using the depth information to design an objective metric for 3D image quality assessment (Zhang and Tam 2005; Hewage et al. 2009). The disparity map, referring to difference in location of an object seen by the left and right eyes, could be generated in many existing stereo matching algorithms. Research on stereo matching tasks has been a hot topic of stereo image processing for decades. More related information about stereo matching could be found in some related literatures (Klaus et al. 2006), which would not be introduced in this book detailly. In general, there are some statistical features generated from the distorted disparity maps to assist to evaluate the image quality of stereopairs.

1. Mean value of disparity: $\mu = E[D]$.
2. Median value of disparity: $median(D)$.
3. Standard deviation of disparity: $\sqrt{E[(D-\mu)^2]}$.
4. Kurtosis of disparity: $E[(D-\mu)^4]/(E[(D-\mu)^2])^2$.
5. Skewness of disparity: $E[(D-\mu)^3]/(E[(D-\mu)^2])^{3/2}$.

When using a Laplacian operator on the disparity map, the differential disparity (δD) is obtained as the feature map. Thus, the statistical value (i.e., mean, standard deviation, kurtosis and skewness) also can be computed by the same methods from the differential disparity map.

1. Mean differential disparity: $\mu_d = E[\delta D]$.
2. Differential disparity standard deviation: $\sqrt{E[(\delta D - \mu_d)^2]}$.
3. Kurtosis of differential disparity: $E[(\delta D - \mu_d)^4]/(E[(\delta D - \mu_d)^2])^2$.
4. Skewness of differential disparity: $E[(\delta D - \mu_d)^3]/(E[(\delta D - \mu_d)^2])^{3/2}$.

In addition, for FR IQA models, there are several computing methods for feature similarities between reference and distorted disparity maps, as list in below (Zhang et al. 2012):

1. Square of the difference $(D - D_{ref})^2$.
2. Absolute error $|D - D_{ref}|$.
3. Square error $1 - \sqrt{D^2 - D_{ref}^2}/255$.

After obtaining similarity maps between D and D_{ref}, statistical features could be extracted by adopting the above introduced methods, or directly integrated into the pooling stage to obtain the final subjective score.

Another way to deal with disparity perception was directly considering disparity maps as feature maps, and some classical IQA methods could be utilized to obtain the quality-aware features from these disparity maps, such as PSNR, SSIM (Wang et al. 2004), MSSIM and so on. Then these disparity maps and other quality-aware feature maps were combined and mapped into the final 3D quality score (Benoit et al. 2008).

In order to make readers better understand the above methods, two SIQA algorithms employing different disparity map generation methods are introduced in the following part. Two different disparity computation algorithms have been selected for these methods: the one described in Felzenszwalb and Huttenlocher (2006), namely, "bpVision" and the one presented in Kolmogorov and Zabih (2002), namely, "kz1". Both of the algorithms modeled the disparity by means of Markov random field (MRF). For the proposed SIQA algorithms, the quality of the distorted stereopair could be measured by the following: (1) The difference between original images and the corresponding distorted images. The usual 2D perceptual quality metrics were applied in this step. (2) The difference between the disparity map of the original stereopair and the distorted stereopair. Perceptual-based distortion metrics could not be applied in this step for disparity maps were not natural images (Kolmogorov and Zabih 2002; Felzenszwalb and Huttenlocher 2006).

The framework of the first approach is shown in Fig. 4.1 (Campisi et al. 2007), which measured the global disparity distortion and then combined this information with the mean evaluation value of the stereopair. Similar to the contents has described before, the quality of the left and right views could be evaluated by SSIM or C4 metrics.

The global disparity distortion measure D_{dg} could be calculated by using the correlation coefficient between the original disparity maps and the corresponding

4.3 SIQA Algorithms Employ the Disparity Information

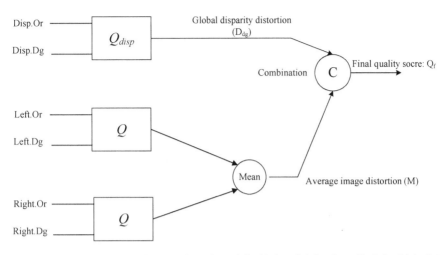

Fig. 4.1 Quality estimation of stereopairs using original left and right views (Left.Or, Right.Or) compared with the degraded versions (Left.Dg, Right.Dg) and the related original disparity map compared to the degraded disparity map (Disp.Or and Disp.Dg) using a global approach

disparity maps processed after image degradation. The final quality score Q_f was obtained after the fusion of the averaged left and right image distortion measures M, and the disparity distortion measure D_{dg}. Both measures M and D_{dg} ranged from 0 (maximum error measure) to 1 (no error measured). Two different fusion rules of M and D_{dg} could be seen in (4.4) and (4.5)

$$d_1 = M \cdot \sqrt{D_{dg}} \qquad (4.4)$$

$$d_2 = M \cdot (1 + D_{dg}) \qquad (4.5)$$

$$d_3 = D_{dg} \qquad (4.6)$$

The main goal was not to determine the best possible combination, but to find out a tradeoff tendency. Main differences between the selected combinations were related to the weight assigned to disparity distortions compared to intra-image (left or right) distortions: d_1 and d_2 combined both disparity and intra-image distortions, but d_3 only considered the disparity distortion. The metric d_1 limited the effect of the disparity distortion measure while d_2 gave more weight to this measure.

The framework of the second approach is shown in Fig. 4.2, the disparity distortion was measured locally and then combined with the monocular quality scores obtained by applying classical 2D IQA algorithms to the left or right image independently. The final quality score was the mean value of left and right distortion measures. Note that, only SSIM could be used in this method, because SSIM measures were available

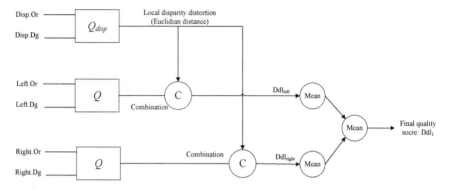

Fig. 4.2 Quality estimation of stereopairs using original left and right views (Left.Or, Right.Or) compared with the degraded versions (Left.Dg, Right.Dg) and the related original disparity map compared to the degraded disparity map (Disp.Or and Disp.Dg) using a local approach

for each pixel of the images by using the SSIM map. C4 was not appropriate for its algorithm focused on discrete areas on the image.

In the second approach, an enhancement of the metric was adapted by using the local SSIM metric in conjunction with the local disparity distortions measures. Generally speaking, SSIM estimated image quality by three factors: luminance, contrast, and structure constancy. In this approach, the contribution of a fourth factor related to the disparity distortion measure was added, which was related to disparity constancy. Following this idea, this approach utilized the Euclidian distance to measure the disparity distortion locally and a weight for the local measure could be obtained (no distortion gives 1, while the maximum distortion measure gives 0). The proposed metric was evaluated by measuring the local SSIM measure map M_{map} and by fusing it with the local disparity distortion measure using point-wise product. The evaluated disparity distortion measure for each pixel p for the left view was (same as the right view)

$$Ddl_{left}(p) = M_{map\_left}(p)\left(1 - \frac{\sqrt{Disp.Or(p)^2 - Disp.Dg(p)^2}}{255}\right). \quad (4.7)$$

The final quality score Ddl_1 was computed by

$$Ddl_1 = \frac{1}{2}\left(\frac{1}{N}\sum_N Ddl_{left}(p) + \frac{1}{N}\sum_N Ddl_{right}(p)\right). \quad (4.8)$$

To further clarify the importance of depth information in stereoscopic image quality assessment, a no-reference SIQA algorithm for JPEG coded stereoscopic images based on local features of distortions and disparity is introduced in detail. This algorithm made use of the 3D depth perception which was strongly dependent on object, structure or texture edge of stereo image content. The block diagram of

4.3 SIQA Algorithms Employ the Disparity Information

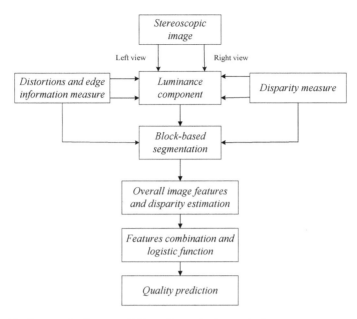

Fig. 4.3 The framework of proposed NR quality evaluation method

the proposed method is shown in Fig. 4.3. Here, blockiness of a block was calculated as the average absolute difference around the block boundary, and zero-crossing technique was used as an edge detector. For simplicity, only the luminance component was added to make overall quality prediction of color stereo images.

The process of measuring image distortions could be summarized as follows. Firstly, the blockiness and zero-crossing of each 8×8 block of the stereo image pair should be calculated separately (left and right views). Secondly, the block-based segmentation algorithm was applied to the left and right images separately to classify edge, and non-edge blocks in the images (Akhter et al. 2010). Thirdly, each value of blockiness and zero-crossing were averaged respectively for edge and non-edge blocks of each image of the stereo pair. Fourthly, for edge blocks and non-edge blocks, the total blockiness and zero-crossing were estimated according to the higher blockiness value and lower zero-crossing value between left and right images respectively. Finally, these blockiness and zero-crossing values were updated by some weighting factors optimized by an optimization algorithm.

Blockiness and zero-crossing measures within each block of the images should be computed firstly in horizontal direction. Supposed image signal be $x(m, n)$ for $m \in [1, M]$ and $n \in [1, N]$, a differencing signal along each horizontal line was calculated by

$$d_h(m, n) = x(m, n+1) - x(m, n), \quad n \in [1, N-1], \; m \in [1, M] \qquad (4.9)$$

Blockiness of a 8 × 8 block in horizontal direction was estimated by

$$B_{bh} = \frac{1}{8}\sum_{i=1}^{8}|d_h(i,8j)| \tag{4.10}$$

where i and $8j$ were the number of row and column position, and $j = 1, 2, \ldots, (N/8)$.
To compute horizontal zero-crossing:

$$d_{h-sign}(m,n) = \begin{cases} 1 & \text{if } d_h(m,n) > 0, \\ -1 & \text{if } d_h(m,n) < 0, \\ 0 & \text{otherwise} \end{cases} \tag{4.11}$$

$$d_{h-mul}(m,n) = d_{h-sign}(m,n) \times d_{h-sign}(m,n+1) \tag{4.12}$$

$$z_h(m,n) = \begin{cases} 1 & \text{if } d_{h-mul}(m,n) < 0, \\ 0 & \text{otherwise} \end{cases} \tag{4.13}$$

where $n \in [1, N-2]$, the size of $z_h(m,n)$ is $M \times (N-2)$. The horizontal zero-crossing of a block (8 × 8) was calculated as follows:

$$ZC_{bh} = \sum_{i=1}^{8}\sum_{j=1}^{8} z_h(i,j) \tag{4.14}$$

The vertical features of blockiness (B_{bv}) and zero-crossing (ZC_{bv}) of the block could be calculated in the same ways. Then the overall features B_b and ZC_b per block were given by

$$B_b = \frac{B_{bh} + B_{bv}}{2}, \quad ZC_b = \frac{ZC_{bh} + ZC_{bv}}{2} \tag{4.15}$$

The average blockiness value of edge, and non-edge areas of the left image were calculated by

$$Bl_e = \frac{1}{N_e}\sum_{b=1}^{N_e} B_{be}, \quad Bl_n = \frac{1}{N_n}\sum_{b=1}^{N_n} B_{bn} \tag{4.16}$$

where N_e and N_n are the number of edge and non-edge blocks of the image. In the same way, the average blockiness values of the right view could be calculated, noted as Br_e and Br_n.

Similarly, the average zero-crossing values of ZCl_e and ZCl_n for the left image, ZCr_e and ZCr_n for the right image could be calculated. The total blockiness features of the stereo image only considered the higher values between the left and right images:

4.3 SIQA Algorithms Employ the Disparity Information

$$B_{e/n}(Bl, Br) = \max(Bl, Br) \tag{4.17}$$

On the contrary, the lower values between the left and right views for average zero-crossing values were selected:

$$ZC_{e/n}(ZCl, ZCr) = \min(ZCl, ZCr) \tag{4.18}$$

Finally, the overall blockiness and zero-crossing of stereo image pair could be obtained by

$$B = B_e^{w_1} \cdot B_n^{w_2} \tag{4.19}$$

$$Z = ZC_e^{w_3} \cdot ZC_n^{w_4} \tag{4.20}$$

where w_1 and w_2 are the weighting factors for the blockiness of edge and non-edge areas; w_3 and w_4 are the weighting factors for the zero-crossing of edge and non-edge areas.

Next the relative disparity should be measured. The principle of the disparity estimation was to divide the left image into 8×8 nonoverlapping blocks, which were edge blocks and non-edge blocks. For each 8×8 block of the left image, stereo correspondence searching was performed according to difference zero-crossing (MDZC) rate between the corresponding block and up to ± 128 pixels of the right image (McKinnon 2009). The disparity estimation approach is shown in Fig. 4.4. "1" and "0" indicated zero-crossing (edge) and nonzero-crossing (non-edge) points respectively.

To measure the relative disparity, firstly, the segmentation algorithm was applied to left image only to classify edge and non-edge blocks. Secondly, block-based difference zero-crossing (DZC) was estimated in the two corresponding blocks between the left and right images (McKinnon 2009). Thirdly, the *DZC* rate values were averaged separately for edge and non-edge blocks. Finally, the values were updated with some weighting factors. The *DZC* of the block could be estimated by

$$DZC = ZCL \oplus ZCr \tag{4.21}$$

where "$\oplus$" indicates a logical Exclusive-OR operation. *DZC* rate (DZCR) was computed by

$$DZCR = \frac{1}{8 \times 8} \sum DZC \tag{4.22}$$

For horizontal direction, let ZCl_h, and ZCr_h be the zero-crossing of a block of left image and the corresponding searching block of right image in horizontal direction, the DZC_h could be calculated by

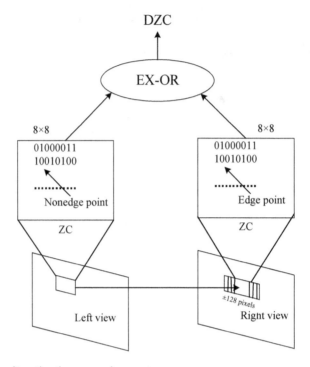

Fig. 4.4 Disparity estimation approach

$$DZC_h = ZCl_h \oplus ZCr_h \tag{4.23}$$

Then $DZCR_h$ of the 8 × 8 block could be calculated by

$$DZCR_h = \frac{1}{8 \times 8} \sum DZC_h \tag{4.24}$$

After that, the average $DZCR_h$ for edge and non-edge blocks of the left image were

$$AZC_{h_e} = \frac{1}{N_e} \sum_{e=1}^{N_e} DZCR_{h_e} \tag{4.25}$$

$$AZC_{h_n} = \frac{1}{N_n} \sum_{e=1}^{N_n} DZCR_{h_n} \tag{4.26}$$

where N_e and N_n are the number of edge and non-edge blocks of the left image.

For vertical direction, $AZCv_e$ and $AZCv_n$ could be calculated in the same way as horizontal direction. Subsequently, the total relative disparity features for edge and non-edge areas were estimated by

4.3 SIQA Algorithms Employ the Disparity Information

$$AZC_e = \frac{AZC_{h_e} + AZC_{v_e}}{2} \quad (4.27)$$

$$AZC_n = \frac{AZC_{h_n} + AZC_{V_n}}{2} \quad (4.28)$$

Finally, the overall relative disparity feature was obtained by

$$DZ = AZC_e^{w_5} \cdot AZC_n^{w_6} \quad (4.29)$$

where w_5 and w_6 are the weighting factors of the disparity features for edge and non-edge areas, respectively.

After all of these works, the artifacts and disparity features could be combined to develop a stereo quality assessment metric. In order to obtain the best suitable features combination equation, the following equations should be investigated:

$$S = \alpha(DZ) \cdot B \cdot Z \quad (4.30)$$

$$S = \alpha + \beta(DZ) \cdot B \cdot Z \quad (4.31)$$

$$S = \alpha(DZ) + \beta(B) + \gamma(Z) \quad (4.32)$$

$$S = \alpha(DZ) + \beta B \cdot Z \quad (4.33)$$

where $\alpha, \beta, \gamma, w_1$ to w_6 are estimated by an optimization algorithm with the subjective test data. In this method, Particle Swarm Optimization (PSO) algorithm was used for optimization (Kennedy and Eberhart 1996). The obtained MOS prediction could be calculated by Parvez Sazzad et al. (2008).

$$MOS_p = \frac{4}{1 + \exp[-1.0217(S-3)]} + 1 \quad (4.34)$$

4.4 The State-of-art SIQA Algorithms Based 2D IQA Weighting

The SIQA algorithm called 3D-MAD (Zhang and Chandler 2015) is introduced in this section, which employed adaptive left/right weighting (Levelt Willem 1965; Blake 2001; Mueller and Blake 1989; Bossink et al. 1993), cyclopean feature images, and machine learning to estimate stereoscopic image quality (Maalouf and Larabi 2011; Chen et al. 2013). This method was proposed to improve the predictive performance of the algorithm on both the symmetrically and asymmetrically distorted stereopairs. There were two main strategies in this work: contrast-weighting strategy and cyclopean-feature-image strategy. Specifically, the 3D-MAD algorithm operated via two main stages to estimate quality of stereoscopic image: (1) 2D-MAD-based quality estimate on both the left and the right views; (2) MCM-based quality estimate on cyclopean feature images (Huang et al. 2010, 2011). Then the two quality estimates were combined through a geometric mean to yield a single value that represented the overall perceived quality degradation of the stereoscopic image.

In the 2D-MAD-based quality estimate stage, the conventional MAD Algorithm was applied on the stereopairs to estimate the perceived distortion corresponding to each monocular view with different trained parameters. Then, the overall 2D-MAD quality was calculated as a weighted sum of two stereo images distortion measures, where the weights were computed based on the normalized block-based contrast. The block diagram of the whole process is shown in Fig. 4.5.

The MAD (Larson and Chandler 2010) algorithm mainly operated based on two strategies: the detection-based strategy and the appearance-based strategy. The former computed the perceived distortion due to visual detection and the latter computed the perceived distortion due to visual appearance dissimilarity. It could be expressed by

$$MAD = (d_{\text{detect}})^{\alpha} \times (d_{appear})^{1-\alpha} \quad (4.35)$$

where $\alpha \in [0, 1]$ serves to adaptively combine the two strategies according to the overall level of distortion.

$$\alpha = \frac{1}{1 + \beta_1 \times (d_{\text{detect}})^{\beta_2}} \quad (4.36)$$

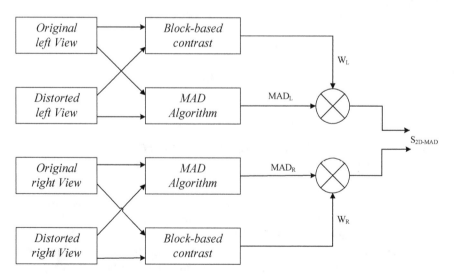

Fig. 4.5 Block diagram of the first 2D-MAD stage

4.4 The State-of-art SIQA Algorithms Based 2D IQA Weighting

where $\beta_1 = 0.369$ and $\beta_2 = 0.153$, which were obtained after training on the 2D LIVE image database. Then each 2D MAD score was weighted by using block-based contrast maps.

Specifically, block-based contrast map was computed in the lightness domain by dividing the image into blocks of 16×16 pixels (with 75% overlap between neighboring blocks) and then the root mean squared contrast of each block was measured.

$$C(b) = \tilde{\sigma}(b)/\mu(b) \qquad (4.37)$$

where $\tilde{\sigma}$ represents the minimum standard deviation among the four 8×8 subblocks within b, $\mu(b)$ represents the average luminance value of block b. Then the block-based contrast maps of the reference left views, the reference right views, the distorted left views and the distorted right views could be computed respectively, which were denoted as C_L^{REF}, C_R^{REF}, C_L^{DST} and C_R^{DST}. The normalized weights of distorted stereopairs were computed via

$$W_L = (\overline{C}_L^{DST}/\overline{C}_L^{REF})^\gamma \qquad (4.38)$$

$$W_R = (\overline{C}_R^{DST}/\overline{C}_R^{REF})^\gamma \qquad (4.39)$$

where $\overline{C}$ is the mean value of C, and $\gamma = 2$ is a factor that is aimed to emphasize higher contrast values. Finally, the 2D-MAD score could be computed by

$$S_{2D-MAD} = \frac{W_L \cdot \exp\left(\frac{MAD_L}{100}\right) + W_R \cdot \exp\left(\frac{MAD_R}{100}\right)}{W_L + W_R} \qquad (4.40)$$

where MAD_L and MAD_R denote the MAD quality scores of the left and right views of the distorted stereoscopic image.

In the second stage, the cyclopean lightness distance and pixel-based contrast maps were used to compute the statistical difference between the reference and distorted stereoscopic images (Vu and Chandler 2011). The block diagram of this stage is shown in Fig. 4.6.

Here, the global lightness distance was computed to measure how much the lightness of each pixel differs from the average lightness of the whole image and the local lightness distance was calculated to measure how much the lightness of each pixel differs from the average lightness of a local area around that pixel. And the pixel-based contrast was used to measure how contrast changed within a local area. These three features could be calculated as

$$f_1(x, y) = \left|L^*(x, y) - \overline{L}_I^*\right| \qquad (4.41)$$

$$f_2(x, y) = \left|L^*(x, y) - \overline{L}_B^*(x, y)\right| \qquad (4.42)$$

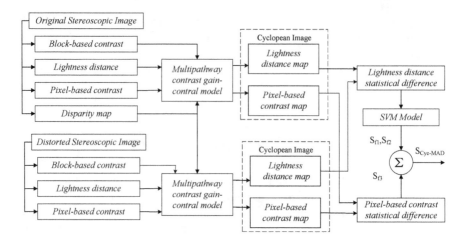

Fig. 4.6 Block diagram of MCM-based cyclopean IQA

$$f_3(x, y) = \frac{L^*(x, y)}{\overline{L}_B^*(x, y) + K} \tag{4.43}$$

where L^* denotes the lightness component in the Commission International de i' Eclairage (CIE) 1976 (L^*, a^*, b^*) color space (CIELAB); $\overline{L}_I^*$ denotes the average lightness value for the whole image; $\overline{L}_B^*(x, y)$ denotes the average lightness value of a 9×9 block centered around pixel (x, y); and $K = 0.001$ is a small constant that prevents division by zero.

Then the disparity-compensated cyclopean maps were built up corresponding to the given lightness distance and pixel-based contrast maps. Firstly, a disparity map of the reference stereopairs (denoted by D) could be calculated by using the segment-based stereo matching approach proposed by Klaus et al. (2006). Secondly the cyclopean feature image could be synthesized based on the MCM. Specifically, there were three gain-control mechanisms in MCM model: (1) attenuation of signal contrast in the non-dominant eye; (2) stronger direct inhibition from the dominant eye; (3) stronger indirect inhibition from the dominant eye to the gain control signal coming from the non-dominant eye. Let C_L and C_R denote the scene contrast corresponding to the left and right eye, separately. By assuming that the left eye was the dominant eye, then after contrast gain control, the signal strengths perceived by the two eyes were modeled as:

$$C'_L = C_L \frac{1}{1 + \frac{\varepsilon_R}{1+\beta\varepsilon_L}}, \quad C'_R = \eta C_R \frac{1}{1 + \frac{\alpha\varepsilon_L}{1+\varepsilon_R}} \tag{4.44}$$

where $\varepsilon_L = \rho(C_R)^{\gamma_1}$, $\varepsilon_R = \rho(\eta C_R)^{\gamma_1}$, ρ is the contrast gain-control efficiency, η is used to model contrast attenuation in right eye; α and β are used to model the stronger inhibition to the right eye from the left eye, γ_1 is the transducer nonlinearity in the

4.4 The State-of-art SIQA Algorithms Based 2D IQA Weighting

gain-control pathway. The perceived contrast of the cyclopean image was given by

$$C = \left[\left(C_L \frac{1}{1 + \frac{\varepsilon_R}{1+\beta\varepsilon_L}} \right)^{\gamma_2} + \left(\eta C_R \frac{1}{1 + \frac{\alpha\varepsilon_L}{1+\varepsilon_R}} \right)^{\gamma_2} \right]^{\frac{1}{\gamma_2}} \quad (4.45)$$

where γ_2 is the transducer nonlinearity for the power summation in binocular contrast combination. According to previous paper (Huang et al. 2011), the perceived cyclopean lightness distance and pixel-based contrast images for each pixel (x, y) were computed by

$$C_i(x, y) = \frac{\left[\left(\eta_L f_{i,L}(x - d_{x,y}, y) \frac{1}{1 + \frac{\varepsilon_R(x-d_{x,y},y)}{1+\beta\varepsilon_L(x-d_{x,y},y)}} \right)^{\gamma_2} + \left(\eta_R f_{i,R}(x, y) \frac{1}{1 + \frac{\alpha\varepsilon_L(x,y)}{1+\varepsilon_R(x,y)}} \right)^{\gamma_2} \right]^{\frac{1}{\gamma_2}}}{\left[\left(\frac{\eta_L}{1 + \frac{\varepsilon_R(x-d_{x,y},y)}{1+\beta\varepsilon_L(x-d_{x,y},y)}} \right)^{\gamma_2} + \left(\frac{\eta_R}{1 + \frac{\alpha\varepsilon_L(x,y)}{1+\varepsilon_R(x,y)}} \right)^{\gamma_2} \right]^{\frac{1}{\gamma_2}}}$$

$(i = 1, 2, 3) \quad (4.46)$

where C_i ($i = 1, 2, 3$) denotes the computed cyclopean lightness distance and pixel-based contrast image; $d_{x,y} = D(x, y)$ denotes an estimated disparity index in D; $\varepsilon_L(x, y) = \rho[\eta_L C_L(x, y)]^{\gamma_1}$; $\varepsilon_R(x, y) = \rho[\eta_R C_R(x, y)]^{\gamma_1}$; C_L and C_R denote the block-based contrast value computed by Eq. 4.37 for the left and right view of a stereo image.

According to the experiment results, these parameter values could be set as: $\gamma_1 = 1.5$, $\rho = 10$. Set $\eta_L = \eta_R = 1$ for the reference stereopairs; For the distorted stereopairs, if the left eye dominates, set $\eta_L = 1$, $\eta_R = 0.9$, otherwise set $\eta_L = 0.9$, $\eta_R = 1$ (Hou et al. 2013). For simplicity, set $\gamma_2 = 0.5$, $\alpha = \beta = 1$.

Based on the cyclopean feature images computed from the lightness distance and pixel-based contrast maps, local statistical differences between the cyclopean feature images for the reference and distorted stereopairs were used to estimate quality by the local standard deviation, skewness, and kurtosis. In this method, three statistical difference values were trained by support vector machine (SVM) learning. The first step was

$$\eta_{\Phi_s}(b) = \sum_{o=1}^{4} \left| \Phi_{s,o}^{ref}(b) - \Phi_{s,o}^{dst}(b) \right| \quad (4.47)$$

where Φ represents the three statistics. Then each statistical difference map was collapsed into a single scalar

$$\overline{\Phi}_s = \left[\frac{1}{B_S} \sum_b \eta_{\Phi_s}(b)^2 \right]^{\frac{1}{2}} \quad (4.48)$$

where B_s denotes the total number of blocks within one statistical difference map (Φ) at a specific scale s (s $\in$ {1, 2, 3, 4, 5}). Thus, for one distorted cyclopean feature image, a vector of 15 statistical-difference-based features was denoted by

$$v = [\bar{\sigma}_1, \bar{\sigma}_2, \bar{\sigma}_3, \bar{\sigma}_4, \bar{\sigma}_5, \bar{\varsigma}_1, \bar{\varsigma}_1, \bar{\varsigma}_1, \bar{\varsigma}_1, \bar{\varsigma}_1, \bar{\kappa}_1, \bar{\kappa}_2, \bar{\kappa}_3, \bar{\kappa}_4, \bar{\kappa}_5] \quad (4.49)$$

Given the feature vectors extracted from the cyclopean feature images, each vector was mapped into a quality score by using the trained SVM models. Specifically, LIBSVM package was used to implement the training. Noted that the method based on SVM was applied only to the lightness distance (local and global) cyclopean images, and two computed feature vectors v_1 and v_2 were obtained. The two corresponding quality scores were denoted as S_1 and S_2. Additionally, the pixel-based contrast statistical features were applied to improve the performance of 3D-MAD on asymmetrically distorted images (Larson and Chandler 2010). Specifically, firstly used the middle radial frequency (third scale) log-Gabor filters to decompose the cyclopean pixel-based contrast images, and then computed the statistical difference map by

$$\eta_{f_3}(b) = \sum_{o=1}^{4} \left[\left| \sigma_{3,o}^{ref}(b) - \sigma_{3,o}^{dst}(b) \right| + 2\left| \xi_{3,o}^{ref}(b) - \xi_{3,o}^{dst}(b) \right| + \left| \kappa_{3,o}^{ref}(b) - \kappa_{3,o}^{dst}(b) \right| \right]$$
$$(4.50)$$

where $\bar{\sigma}$, ξ, and κ denote the three statistics respectively. Then the final scalar value of contrast distortion was given by

$$S_3 = \left[\frac{1}{B} \sum_b \eta_{f_3}(b)^2 \right]^{\frac{1}{2}} \quad (4.51)$$

where B denotes the total number of blocks within the pixel-based contrast statistical difference map.

Finally, the MCM-based cyclopean feature image quality was calculated by

$$S_{Cyc-MAD} = \frac{S_1 + S_2}{10} + S_3 \quad (4.52)$$

The final step of 3D-MAD algorithm was to combine the 2D-MAD-based estimate and the MCM-based estimate to yield an overall perceived distortion estimate of the stereoscopic image. Specifically, S_{3D-MAD} was calculated as a product of S_{2D-MAD} and $S_{Cyc-MAD}$.

$$S_{3D-MAD} = S_{2D-MAD} \times S_{Cyc-MAD} \qquad (4.53)$$

where the smaller value of S_{3D-MAD} means the better stereoscopic image quality, and a value of $S_{3D-MAD} = 0$ indicates the original image.

4.5 Summary

In this chapter, the development of SIQA based 2D IQA weighting are reviewed in detail. At the earliest stage, 2D image quality assessment methods (e.g. SSIM, C4, UQI and RRIQA) were directly applied on the left and right views respectively and then calculated the mean value of the two views. However, for human visual system, binocular vision is a complex visual process that binocular fusion and binocular rivalry coexist at the same point in space and time. People have made a lot of efforts to imitate HVS like human. At first, the depth/disparity information was employed to judge the stereo image quality, and then, the cyclopean view was proposed, which was defined as the average of the left image and the disparity-compensated right image (Maalouf and Larabi 2011), or synthesized by using a linear model (Chen et al. 2013), in which the coefficients of the linear model were computed based on the local Gabor filter energy. Based on these knowledges, a state-of-art algorithm was presented to evaluate stereoscopic image quality, called 3D-MAD, but the performance of this algorithm still needs to be improved. In the next chapter, SIQA based binocular combination will be introduced.

References

Ahmed S, Larabi MC (2014) Stereoscopic 3D image quality assessment based on cyclopean view and depth map. In: International conference on consumer electronics, Berlin, Germany, pp 335–339

Akhter R, Parvez Sazzad ZM, Horita Y, Baltes J (2010) Noreference stereoscopic image quality assessment. In: Stereoscopic displays and applications XXI, Proceedings of the SPIE-Society of Photo-Optical Instrumentation Engineers, San Jose, CA, USA, p 7524

Benoit A, Callet PL, Campisi P (2008) Using disparity for quality assessment of stereoscopic images. In: IEEE international conference on image processing, San Diego, California, USA, pp 389–392

Blake R (2001) A primer on binocular rivalry, including current controversies. Brain Mind 2(1):5–38

Boev A, Poikela M, Gotchev A, Aksay A (2009) Modelling of the stereoscopic HVS

Bossink CJ, Stalmeier PF, De Weert CM (1993) A test of Levelt's second proposition for binocular rivalry. Vision Res 33:1413–1419

Braddick OJ (1979) Binocular single vision and perceptual processing. Proc Roy Soc Lond Ser B Biol Sci Lond 204(1157):503–512

Campisi P, Callet PL, Marini E (2007) Stereoscopic images quality assessment. In: 15th European signal processing conference, Pozan, Poland, pp 3–7

Carnec M, Le callet P, Barba D (2003) An image quality assessment method based on perception of structural information. In: IEEE international conference on image processing, Barcelona, Spain, pp 185–188

Chen M-J, Su C-C, Kwon D-K, Cormack LK, Bovik AC (2013) Full-reference quality assessment of stereopairs accounting for rivalry. Sig Process Image Commun 28(9):1143–1155

Felzenszwalb PF, Huttenlocher DP (2006) Efficient belief propagation for early vision. Int J Comput Vision 70(1):41–54

Goldmann L, Ebrahimi T (2010) 3D quality is more than just the sum of 2D and depth. In: IEEE international workshop on hot topics in 3D, Singapore

Gorley P, Holliman N (2008) Stereoscopic image quality metrics and compression. In: Proceedings of SPIE-Society of Photo-Optical Instrumentation Engineers, p 6803

Hewage CTER, Worrall ST, Dogan S, Kondoz AM (2008) Prediction of stereoscopic video quality using objective quality models of 2-D video. Electron Lett 44(16):963–965

Hewage CTER, Worrall ST, Dogan S (2009) Quality evaluation of color plus depth map-based stereoscopic video. IEEE J Sel Top Sig Process 3(2):304–318

Hou F, Huang C-B, Liang J, Zhou Y-F, Lu Z-L (2013) Contrast gaincontrol in stereo depth and cyclopean contrast perception. J Vision 13(8)

Huang C-B, Zhou J-W, Zhou Y-F, Lu Z-L (2010) Contrast and phase combination in binocular vision. PLoS ONE 5:5075

Huang C-B, Zhou J-W, Zhou Y-F, Lu Z-L (2011) Deficient binocular combination reveals mechanisms of anisometropic amblyopia: signal attenuation and interocular inhibition. J Vis 11:1–17

Kaptein RG, Kuijsters A, Lambooij Marc TM, IJsselsteijn WA, Heynderickx I (2008) Performance evaluation of 3D-TV systems. In: Proceedings of SPIE: image quality and system performance, vol 6808

Kennedy J, Eberhart R (1996) Particle swarm optimization. In: Proceedings of the IEEE international conference on neural networks, Perth, Australia, pp 1942–1948

Klaus A, Sormann M, Karner K (2006) Segment-based stereo matching using belief propagation and a self-adapting dissimilarity measure. In: Proceedings of the 18th international conference on pattern recognition, vol 3, pp 15–18

Kolmogorov V, Zabih R (2002) Multi-camera scene reconstruction via graph cuts. In: Proceedings of the 7th European conference on computer vision, Copenhagen, Denmark, pp 82–96

Lambooij M, IJsselsteijn W, Fortuin M, Heynderickx I (2009) Visual discomfort and visual fatigue of stereoscopic display: a review. J Imaging Sci Technol 53(3):030201/114

Larson EC, Chandler DM (2010) Most apparent distortion: full reference image quality assessment and the role of strategy. J Electron Imaging 19(1):011006

Levelt Willem JM (1965) On binocular rivalry. Institute for Perception RVO-TNO, National Defence Research Organization TNO, Soesterberg, Netherlands

Maalouf A, Larabi MC (2011) CYCLOP: a stereo color image quality assessment metric. In: IEEE international conference on acoustics, speech and signal processing, pp 1161–1164

McKinnon BP (2009) Point, line segment, and region-based stero matching for mobile robotics. Department of Computer Science, University of Manitoba

Mueller TJ, Blake R (1989) A fresh look at the temporal dynamics of binocular rivalry. Biol Cybern 61:223–232

Parvez Sazzad ZM, Kawayoke Y, Horita Y (2008) No reference image quality assessment for JPEG2000 based on spatial features. Sig Process Image Commun 23(4):257–268

Qi F, Zhao D, Gao W (2015) Reduced reference stereoscopic image quality assessment based on binocular perceptual information. IEEE Trans Multimedia 17(12):2338–2344

Vu C, Chandler DM (2011) Main subject detection via adaptive feature refinement. J Electron Imaging 20(1):013011

Wang Z, Bovik AC (2002) A universal image quality index. IEEE Sig Process Lett 9(3):81–84

References

Wang Z, Simoncelli EP (2005) Reduced-reference image quality assessment using a wavelet-domain natural image statistic model. In: Human vision and electronic imaging X, Proceedings of the SPIE-Society of Photo-Optical Instrumentation Engineers, vol 5666, pp 149–159

Wang Z, Bovik AC, Sheikh HR, Simoncelli EP (2004) Image quality assessment: from error visibility to structural similarity. IEEE Trans Image Process 13(4):600–612

Yasakethu SLP, Hewage CTER, Fernando WAC, Kondoz AM (2008) Quality analysis for 3D video using 2D video quality models. IEEE Trans Consum Electron 54(4):1169–1176

Zhang L, Tam WJ (2005) Stereoscopic image generation based on depth images for 3D TV. IEEE Trans Broadcast 51(2):191–199

Zhang L, Zhang L, Mou X (2012) A comprehensive evaluation of full reference image quality assessment algorithms. In: International conference on image processing, Lake Buena Vista, Florida, USA, pp 1477–1480

Zhang Y, Chandler DM (2015) 3D-MAD: a full reference stereoscopic image quality estimator based on binocular lightness and contrast perception. IEEE Trans Image Process 24(11):3810–3825. https://doi.org/10.1109/TIP.2015.2456414

Chapter 5
Stereoscopic Image Quality Assessment Based on Binocular Combination

Abstract Only employing depth information in stereoscopic image quality assessment models cannot simulate human visual characteristics well. Thus, a cyclopean image generated from the left and right views is designed to overcome this defect. Different methods to generate a cyclopean image are discussed firstly in this chapter. Then two region classification strategies to deal with no-matched pixels caused by different angles of two views are introduced. Finally, visual fatigue and visual discomfort prediction models are developed to simulate the negative influence of non-corresponding areas in stereo pairs.

Keywords Stereoscopic image quality assessment · Cyclopean image · Visual fatigue · Visual discomfort

5.1 Introduction

In the previous chapter we have explored the Stereo Image Quality Assessment (SIQA) methods based on two-dimensional (2D) IQA metrics. These methods can be classified into two categories according to whether depth information is used or not. The straightforward way to estimate stereoscopic image quality is that directly apply 2D quality metric on the left view and right view, and then generate the overall quality scores (Yasakethu et al. 2008; Gorley and Holliman 2008). However, this kind of methods is not accurate in most cases, for that it completely ignores the performance of Human Visual System (HVS). Obviously, even if the depth information is utilized in this kind of method, it cannot model human visual characteristics well, for human visual perception is a complex visual process, which includes binocular fusion, binocular suppression and other factors. Therefore, a cyclopean image generated from the left and right views is designed to model these complex properties. The invention of the cyclopean image provides an important clue to solve the SIQA problem, as the result of combination of locally matched stereo regions considering human perception properties such as binocular fusion and binocular rivalry. The earliest method is shown in Henkel (1997), specifically, for each pixel, they took a

(a) (b) (c)

Fig. 5.1 **a**, **b** are the left and right views from certain stereopair, **c** is the corresponding cyclopean image

local window from the left view and average it with the disparity-matched window from the right view. The concrete example is shown in Fig. 5.1.

After obtaining the cyclopean image, the state-of-art 2D IQA metrics can be applied to the generated image. This binocular combination model is just too simple to model human vision system. Inspired by the idea of this method, some advanced methods to form the cyclopean through different binocular combination were proposed. Researchers found that the energy response of Gabor filter can model stimulus strength and simulate binocular rivalry. Therefore, many models have been constructed based on Gabor filter. Disparity information is another important clue to evaluate the quality of stereoscopic images, thus many stereo algorithms have been designed to calculate the estimated disparity map. Considering the different angle of two views, the no-matched pixels may exist in some regions of the left view and right view. These non-corresponding areas will produce some negative influence in the process of binocular fusion. Last but not least, visual fatigue and visual discomfort are also closely connected to the viewing experience of human, which may result in the descent of overall quality scores.

In Sect. 5.2, we will first introduce some typical vision models, such as eye-weighting model, vector summation model, neural network model and gain-control theory model. Then some SIQA algorithms based on these models and some improvements of typical methods will be discussed. In Sect. 5.3, two region classification strategies to deal with no-matched pixels caused by different angles of two views will be introduced. In Sect. 5.4, visual fatigue and visual discomfort prediction models will be constructed, in which non-corresponding areas will produce some negative influence in the process of binocular fusion.

5.2 How to Form the Cyclopean Image

The main challenges of SIQA based binocular combination come from the interactions between the two eyes, so a deep understanding of the physiological researches (Blake and Wilson 2011; Steinman et al. 2000; Howard and Rogers 2008; Mather 2008) of binocular vision is helpful to build the effective computational models for stereopairs. The binocular combination for matched stereopairs will be introduced in detail in this section.

Studies of human vision have shown that our two eyes see the world from a slightly different angle. Disparity caused by different viewing angles of two eyes is the basis of constructing the cyclopean image. Binocular combination refers to combining the left and right views on the retinas to generate a single cyclopean image. Important clues about binocular combination can be obtained from the data about the perceived luminance of surface regions under the condition of binocular observation, including data in response to the sum of binocular luminance of Ganzfelds (Bolanowski 1987; Bourassa and Rule 1994), the U-shaped data of Fechner's paradox that violates binocular brightness summation, as well as the effects of different combinations of monocular and binocular contours and surface luminance differences on threshold sensitivity to monocular flashes of light (Cogan 1982). How to coordinate these contradictory data properties has been a huge challenge for building a binocular fusion model. Many researchers have exerted great efforts to quantitatively simulate these properties by developing corresponding vision models, which are generally divided into four types: eye-weighting models, vector summation models, neural network models, and gain-control theory models (Ding and Sperling 2006). These models can be briefly introduced as follows.

Eye-weighting models compute the cyclopean image by summing the left and right views linearly, which can be described as (Engel 1967).

$$C = \left[(w_L I_L)^\beta + (w_R L_R)^\beta\right]^{\frac{1}{\beta}} \quad (5.1)$$

where C is the generated cyclopean image; w_L and w_R are the weighting factors of the left and right views, respectively; $\beta = 1$ or 2 refers to monocular luminance or quadratic luminance; I_L and I_R represent the left and right images, respectively. Even though some other models with different weighting factors were proposed, the existing model is simple enough and the quadratic of luminance can explain the binocular combination behavior when the input luminance is symmetric for both eyes. However, as a result of neglecting the interaction between two eyes, these models failed in matching with Fechner's paradox and explaining cyclopean perception.

Considering these defects of eye-weighting models, some researchers proposed vector summation models, which generate the binocular brightness perception by summing two orthogonal vectors with some normalization (Schrodinger 1926; MacLeod 1972). The typical vector summation model (Curtis and Rule 1978) can be given by

$$C(x, y) = \sqrt{E_L^2(x, y) + E_R^2(x + d, y) + 2 \cdot E_L(x, y) E_R(x + d, y)} \qquad (5.2)$$

where E_L and E_R are the luminance of left eye and right eye, respectively.

Neural network models firstly compute the neural response of each eye, and successfully achieve Fechner's paradox using neural thresholds and asymmetric excitation and inhibition. The left and right inputs can be expressed by

$$N_L = f(E_L - h_R E_R), \quad N_R = f(E_R - h_L E_L) \qquad (5.3)$$

where E_L and E_R are monocular luminance signals for left and right eyes, respectively; h_L and h_R control the inhibitory signals or threshold characteristics. Obviously, N_L and N_R are the responses of cells receiving strong excitation to one eye and weak inhibition from the other eye. Finally, these responses are combined by

$$C = N_L + N_R \qquad (5.4)$$

Of course, there are some other neural network models proposed by researchers. For example, Lehky's nonlinear summation model is most similar to how the Form-And-Color-And-Depth (FACADE) combines monocular brightness signals (Grossberg 1987, 1994, 1997; Lehky 1983). Cogan's two channel model firstly use separate monocular and binocular channels (Cogan 1987). Anderson and Movshon's distribution model (Anderson and Movshon 1989) possesses several linear binocular channels, each of which has a degree of ocular dominance wherein each channel is more or less sensitive to each eye.

Gain-control theory model was proposed by relatively recent study (Ding and Sperling 2006). In this model the binocular combination is defined as:

$$C(x, y) = \frac{1 + R_L(x, y)}{1 + R_L(x, y) + R_R(x + d, y)} \cdot E_L(x, y)$$
$$+ \frac{1 + R_R(x + d, y)}{1 + R_L(x, y) + R_R(x + d, y)} \cdot E_R(x + d, y) \qquad (5.5)$$

where E_L and E_R are the luminance of the left and right view, respectively; R_L and R_R are the energy response sums of all the frequency channels for the left and right view, respectively. This model is relatively dependent, which sum the very low-contrast stimuli to the left and right eyes linearly to form the predicted cyclopean image. Meanwhile, both binocular combination and binocular rivalry are taken into account, which coincides with the perceived natural vision.

Then many SIQA algorithms motivated by these binocular models were proposed to evaluate the quality of stereo images. Chen firstly developed Full-Reference (FR) SIQA method for binocular rivalry (Chen et al. 2013). Binocular rivalry is a perceptual effect that happens when the light images on the two retinas are dissimilar. The stimuli from the left and right eyes alternately dominates human vision. Binocular suppression is a special case of binocular rivalry, when no rivalrous fluctuations

5.2 How to Form the Cyclopean Image

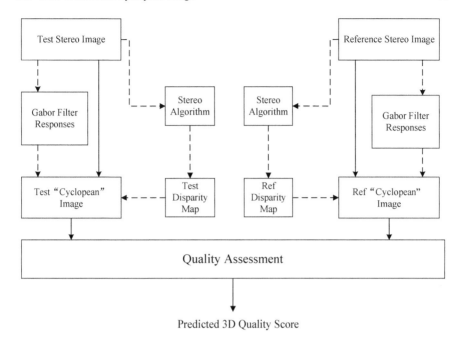

Fig. 5.2 The framework for 3D quality assessment proposed by Chen

happen between two stimuli on the retinas. The framework of Chen's model is shown in Fig. 5.2.

In the framework above, the estimated disparity map is generated by a stereo algorithm, and Gabor filter responses are obtained by using a bandpass filter bank on the stereopairs. Then the cyclopean image can be synthesized from the stereopairs, the estimated disparity map and the Gabor filter responses. Finally, FR 2D IQA methods such as SSIM can be applied to cyclopean images to obtain 3D quality scores.

In Chen's paper, three stereo algorithms were chosen to generate disparity map. The first algorithm is sum-of-absolute difference (SAD), which compute the disparity value of each pixel in stereopairs by minimizing the SAD between this pixel and its horizontal shifted pixels in the other view with ties broken by selecting the lower disparity solution. The second algorithm is segment-based stereo matching using belief propagation and a self-adapting dissimilarity measure (Klaus et al. 2006). The third algorithm generate the disparity map of stereopairs by maximizing the SSIM scores between the left and right views along the horizontal direction. Experiments prove that there is no significant difference in the performance among these algorithms, the third algorithm slightly outperforms others. In addition to these algorithms, some other methods obtained better performance were proposed. In Jian's paper, an optical flow estimation-based algorithm was adopted to generate the disparity map (Sun et al. 2010). In Sun's paper, they spread the model by employing the position shift mechanism instead of the phase shift mechanism of DSKL model to generate the disparity map (Li et al. 2014). In Shao's paper, the estimated disparity map was generated by stereo matching algorithm (Kolmogorov and Zabih 2001) due to its good performance for high-quality stereoscopic images.

Inspired by Levelt's experiments (Levelt 1968), Chen use the energy of Gabor filter bank responses on the left and right images to model the stimulus intensity, and to simulate the binocular rivalry behavior. The Gabor filter bank extracts features from the luminance and chrominance channels. These filters can closely model frequency-orientation decompositions in primary visual cortex and capture energy in a highly localized manner in both space and frequency (Field 1987). A complex 2-D Gabor filter is expressed as

$$G(x, y, \sigma_x, \sigma_y, \zeta_x, \zeta_y, \theta) = \frac{1}{2\pi \sigma_x \sigma_y} e^{-\frac{1}{2}\left[\left(\frac{R_1}{\sigma_x}\right)^2 + \left(\frac{R_2}{\sigma_y}\right)^2\right]} e^{i(x\zeta_x + y\zeta_y)} \quad (5.6)$$

where $R_1 = x\cos\theta + y\sin\theta$, and $R_2 = -\sin\theta + y\cos\theta$, σ_x and σ_y are the standard deviations of an elliptical Gaussian envelope along x and y axes, ζ_x and ζ_y are spatial frequencies, and θ orients the filter. The local energy can be computed by summing Gabor filter magnitude responses over four orientations (horizontal, both diagonals, and vertical) at a spatial frequency of 3.67 cycles/degree. Because the maximum bandwidth of a Gabor filter is limited to approximately one octave, some methods adopted Log-Gabor filter instead of Gabor filter, which discards the DC components and can be used in the condition of arbitrary bandwidth (Fleet et al. 1996). The frequency response of the Log-Gabor filters utilized in SIQA methods can be expressed by

$$G_{s,o}(w, \theta) = \exp\left[-\frac{(\log(w/w_s))^2}{2\delta_s^2}\right] \times \exp\left[-\frac{(\theta - \theta_o)^2}{2\delta_o^2}\right] \quad (5.7)$$

where S and O are spatial scale and orientation index, θ is the orientation angle, δ_s and δ_o control the strength of the filter, w and w_s are the normalized radial frequency and the corresponding center frequency of the filter.

In Chen's model, as mentioned eye-weighting model before, the localized linear model that they used to synthesize a cyclopean image is:

$$CI(x, y) = W_L(x, y) \times I_L(x, y) + W_R((x + d), y) \times I_R((x + d), y) \quad (5.8)$$

where CI is the synthesized cyclopean image, I_L and I_R are the left and right images respectively, and d is the disparity index of pixels from I_L to corresponding pixels in I_R. The weighting factors W_L and W_R are calculated form the normalized Gabor filter magnitude responses, which can be defined by

$$W_L(x, y) = \frac{GE_L(x, y)}{GE_L(x, y) + GE_R((x + d), y)} \quad (5.9)$$

5.2 How to Form the Cyclopean Image

$$W_R(x+d, y) = \frac{GE_R((x+d), y)}{GE_L(x, y) + GE_R((x+d), y)} \quad (5.10)$$

where GE_L and GE_R are the summation of the convolution responses to the left and right images by Eq. 5.6. Through Eqs. 5.9 and 5.10, increased Gabor energy of either stimulus suppresses the representation of the other view, which conforms to the binocular rivalry. Finally, full reference 2D IQA metrics can be applied to the cyclopean images to obtain overall quality scores.

The limitation of Chen's method is that the cyclopean image for the distorted stereopair is synthesized from the poor disparity information and only 2D quality metrics are applied on the cyclopean view. Lin et al. proposed the frequency integrated model in which binocular integration behaviors using a difference of the Gaussian energy-based gain control model, and then 2D IQA metrics are applied to evaluate the monocular view (Lin and Wu 2014). Shao et al. constructed a full reference SIQA model in which the final quality score is obtained by generating a binocular combination using sparse energy and sparse complexity through learning binocular receptive field properties (Shao et al. 2015). After that, Shao et al. proposed a method to evaluate the perception quality of stereo images by simulating the visual properties of the primary visual cortex (Shao et al. 2017). Lin et al. developed a full reference SIQA approach by jointly considering the cyclopean amplitude and cyclopean phase (Lin et al. 2017). Zhang et al. use binocular lightness and contrast perception to evaluate the stereo image quality (Zhang and Chandler 2015). Xu et al. built the SIQA model based on learning manifold color visual properties (Xu et al. 2016). Except for these full reference SIQA methods mentioned above, Zhou et al. proposed blind quality estimator for stereo images based on binocular combination and extreme learning machine (Zhou et al. 2017).

Here, we will focus on a state-of-art FR SIQA algorithm (Sun et al. 2018), which takes the changes of luminance into consideration and uses support vector regression (SVR) to generate final quality scores. They used log-Gabor filter mentioned before to extract multi-scale energy responses for given reference and distorted stereopairs. Furthermore, luminance maps are generated from stereopairs to reflect the quality degradation of stereo images. Then binocular fusion and binocular rivalry models are applied to extract binocular visual features from energy responses and luminance maps. Finally, Local Binary Patterns (LBPs) are adopted as features encoding strategy to generate a set of feature vectors, which are mapped into an overall quality score by the SVR.

The gain-control and gain-enhancement model is adopted in this method, which can explain the neural processing of binocular vision well. The energy response of this model is given by

$$E_v(x, y; f) = \frac{\sum_\theta |C_v(x, y; f, \theta)|}{g_c}, \quad E_v^*(x, y; f) = \frac{\sum_\theta |C_v(x, y; f, \theta)|}{g_e} \quad (5.11)$$

where $C_V(x, y; f, \theta)$ is the energy response map of a monocular view in a specific scale and orientation, in which $v \in \{l, r\}$. E_v and E_v^* are the total energy response across different orientations in different scales. g_c and g_e are the gain-control threshold and the gain-enhancement threshold respectively. Then the monocular energy responses of stereopairs can be expressed by

$$|C_l(x, y; f)|_{ce} = \frac{1 + \frac{E_r^*(x,y;f)}{1+\beta E_l(x,y;f)}}{1 + \frac{E_r(x,y;f)}{1+\alpha E_l(x,y;f)}} \sum_\theta |C_l(x, y; f, \theta)| \quad (5.12)$$

$$|C_r(x, y; f)|_{ce} = \frac{1 + \frac{E_l^*(x,y;f)}{1+\beta E_r(x,y;f)}}{1 + \frac{E_l(x,y;f)}{1+\alpha E_r(x,y;f)}} \sum_\theta |C_r(x, y; f, \theta)| \quad (5.13)$$

where α and β are different gain-control efficiencies to reflect different influence in final monocular energy response. Finally, the monocular energy response can be incorporated into the binocular energy response, which is expressed as follows:

$$|C(x, y; f)| = \left\| |C_l(x, y; f)|_{ce} + |C_r(x, y; f)|_{ce} \right\| \quad (5.14)$$

In the same way, we can denote the binocular energy response of reference and distorted stereopairs as $|C_R(x, y; f)|$ and $|C_D(x, y; f)|$, respectively. Then the similarity map of binocular energy response in different scales for binocular fusion property is defined by

$$S_{BF}(x, y; f) = \frac{2|C_R(x, y; f)| \times |C_D(x, y; f)| + T_1}{|C_R(x, y; f)|^2 + |C_D(x, y; f)|^2 + T_1} \quad (5.15)$$

where T_1 is a constant to avoid the denominator being equal to zero. The binocular luminance map generated from the left and right views is expressed by

$$|I(x, y; f)| = \sqrt{I_l(x, y; f)_{ce}^2 + I_r(x, y; f)_{ce}^2 + 2I_l(x, y; f)_{ce} \times I_r(x, y; f)_{ce}}$$
$$= I_l(x, y; f)_{ce} + I_r(x, y; f)_{ce} \quad (5.16)$$

Only use binocular fusion model cannot conform to the human visual system well, so this paper adopted binocular rivalry model to improve performance at the same time. The similarity map generated by dissimilarity quantification between original and distorted images in monocular views to measure the quality degradation of corresponding distorted images is defined by

$$S_v(x, y; f) = \frac{2|C_{R,v}(x, y; f)| \times |C_{D,v}(x, y; f)| + T_2}{|C_{R,v}(x, y; f)|^2 + |C_{D,v}(x, y; f)|^2 + T_2} \quad (5.17)$$

where $C_{R,v}(x, y; f)$ and $C_{D,v}(x, y; f)$ represent the energy response maps of reference and distortion stereopairs in different scales, where $v \in \{l, r\}$, and T_2 is a constant

5.2 How to Form the Cyclopean Image

to avoid the denominator being equal to zero. Generally, a linear model is used to simulate binocular rivalry, which is defined by

$$S_{BR}(x, y; f) = w_l(x, y; f) \times S_l(x, y; f) + w_r(x, y; f) \times S_r(x, y; f) \quad (5.18)$$

where w_l and w_r are the weighting factors to represent the relative receiving contribution of two eyes, which are expressed by

$$w_l(x, y; f) = \frac{1 + M_l(x, y; f)}{1 + M_l(x, y; f) + M_r(x, y; f)} \quad (5.19)$$

$$w_r(x, y; f) = \frac{1 + M_r(x, y; f)}{1 + M_l(x, y; f) + M_r(x, y; f)} \quad (5.20)$$

where $M_l(x, y; f)$ and $M_r(x, y; f)$ are the energy responses of the left and right views in different scales respectively.

LBPs are always served as an effective structural and textural information operator and used in dimension reduction (Ojala et al. 2002). Specifically, this method can form a non-directional binary local structural pattern by comparing eight local neighbor pixels with a central pixel and concatenating the results. The limitation of this method is that the statistical histogram of LBP maps will output 256-dimensional features, which is too high for calculation. Thus the improved method, namely the rotation-invariant uniform LBPs, was proposed to solve this problem, which only has 10-dimensional outputs for eight neighboring pixels. After features encoding by the rotation-invariant uniform LBPs, the obtained 160-dimensional features are mapped into a final objective quality score by SVR. The whole framework of this method is illustrated in Fig. 5.3.

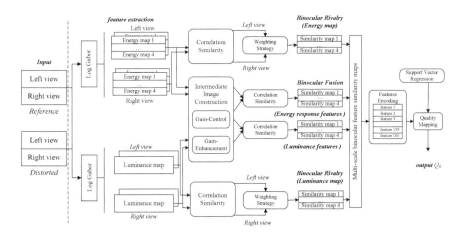

Fig. 5.3 Framework of the SIQA method proposed by Sun

5.3 Region Classification Strategy

In some full-reference SIQA methods, it is assumed that the original stereo images are well calibrated, so that there is no binocular suppression. Generally, if a pixel does not have a matching correspondence in the other view, monocular vision appears on the non-corresponding retinal points. Otherwise, binocular fusion appears on the corresponding retinal points. However, the introduction of different level distortion affects the identification of binocular disparity, and then influence the combination of binocular perception. To eliminate these potential interferences, region classification strategies are adopted in some approaches. Generally, the distorted stereoscopic images are categorized into three kinds of regions, which are non-corresponding region, binocular fusion region and binocular suppression region respectively. Each region is calculated independently based on its own perception property, and then combined into an overall quality score. The detailed process of region classification is described as follows.

In the left and right views of stereopairs, the occlusion/disocclusion inevitably happens in the images for different viewing angle. For instance, the background behind the foreground in the left view will be invisible in the right view, which means the right image provides no information about corresponding areas in the left image. Therefore, we regard the occluded/disoccluded regions in the estimated disparity maps as non-corresponding region. The left non-corresponding region R_{nc}^l in the left views and the right non-corresponding region R_{nc}^r in the right views make up the non-corresponding region of the distorted stereopairs together, which can be denoted as R_{nc} for there is no overlap region.

$$R_{nc} = R_{nc}^l \cup R_{nc}^r \tag{5.21}$$

As mentioned before, binocular suppression happens under the conditions of the contents are mismatched in the stereopairs and the matched contents are not consistent, such as the corresponding regions are matched but with large matching error. Based on the above two conditions, the binocular suppression region can be defined. First, left-right consistency check is applied to detect the inconsistent pixels in the disparity maps. The pixels are regarded as mismatched regions if the sum of the matched disparity values is larger than T_{LR}, which can be expressed by

$$|d_L(x, y) + d_R(x - d_L(x, y), y)| \geq T_{LR} \tag{5.22}$$

where $d_L(x, y)$ and $d_R(x, y)$ denote the left disparity map and the right disparity map, respectively. Even though the above condition is not satisfied for the matched pixels, we need to judge further if the matching error is larger than T_{ME}, which is defined as follows:

$$|I_L(x, y) - I_R(x - d_L(x, y), y)| \geq T_{ME} \tag{5.23}$$

5.3 Region Classification Strategy

where T_{ME} is a threshold determined by binocular just noticeable difference (BJND) model (Zhao et al. 2011). Binocular fusion still works if the inter-difference between two views is smaller than the threshold T_{ME}. Finally, the binocular suppression region of the distorted stereopairs can be described as

$$R_{bs} = \sup(R_{bs}^l, R_{bs}^r) \tag{5.24}$$

where $\sup()$ represents the binocular suppression operation, R_{bs}^l and R_{bs}^r represent the left binocular suppression region and right binocular suppression region respectively. Note that the binocular suppression operation is not a simple superposition of R_{bs}^l and R_{bs}^r, instead it is a complex non-linear neurophysiological process, which includes formation of the cyclopean image (Julesz 1971). Therefore, we usually evaluate the quality of the left suppression and right suppression region independently, and then combine them to generate the overall quality score for the region. Generally, the increase strength of distortion in the stereopairs will lead to more binocular suppression regions.

After excluding the non-corresponding region and binocular suppression region mentioned above, the remaining region in the distorted stereoscopic image is binocular fusion region, which can be described as

$$R_{bf} = fus(R_{bf}^l, R_{bf}^r) \tag{5.25}$$

where $fus()$ represents the binocular fusion operation, R_{bf}^l and R_{bf}^r represent the left binocular fusion region and right binocular fusion region, respectively.

In this method the score of binocular suppression or rivalry region is calculated as the maximum score from the left and right quality scores, which is not in compliance with the human vision system. To optimize this problem, Binocular Just Noticeable Difference (BJND) is used for weighting each region according to its perceptual significance (Ahmed et al. 2014; Hachicha et al. 2013).

In Hachicha's paper, the binocular vision area is divided into four disjoint regions in more detail, which are occlusion region (denoted as O_v), invisible distortion region (denoted as T_v), binocular suppression region (denoted as S_v) and binocular rivalry region (denoted as R_v). These regions in the left view are defined as follows mathematically.

$$O_l = \{(i, j) \in l, d_l(i, j) = 0\} \cup \{(i, j) \in l, i - d_l < 0\} \tag{5.26}$$

$$T_l = \{(i, j) \in \overline{O_l}, \sum_{(p,q) \in B} |I_l(p, q) - I_{\tilde{r}}(p, q)| < \sum_{(p,q) \in B} BJND_l(p, q)\} \tag{5.27}$$

$$S_l = \overline{O_l} \cap \overline{T_l} \cap C_l \cap L_l, \quad R_l = \overline{O_l} \cap \overline{T_l} \cap C_l \cap \overline{L_l} \tag{5.28}$$

$$C_l = \{(i, j) \in l, \sum_{(p,q) \in B} C_{\tilde{l}}(p, q) > \sum_{(p,q) \in B} C_{\tilde{r}}(p - d_l, q)\} \tag{5.29}$$

$$L_l = \{(i,j) \in \overline{O}_l, \sum_{(p,q) \in B} |I_{\tilde{l}}(p,q) - I_{\tilde{r}}(p-d_l,q)| < \sum_{(p,q) \in B} BJND_{\tilde{j}}(p,q)\} \tag{5.30}$$

where $C_v, v \in \{\tilde{l}, \tilde{r}\}$ is the local contrast computed as in Hachicha et al. (2012), B is a square block of size $w \times w$ centered at pixel (i, j), d_l is the disparity value at (i, j) of the left view. In the same way, O_r, T_r, S_r, R_r, L_r and C_r of the right view can be calculated. Note that, the horizontal pixel position is positive shift, i.e. $i + d_r$, and $i + d_r$ should be less than the image width size.

Finally, the overall quality score is derived as follows

$$SM = \sqrt{\frac{1}{N} \sum_{(i,j) \in \{l\} \cup \{r\}} sm(i,j)} \tag{5.31}$$

where N is the cardinality of $O_l \cup O_r \cup C_l \cup C_r$, $sm(i, j)$ is expressed by

$$sm(i,j) = \begin{cases} \Delta m_l(i,j)^2 & \text{if } (i,j) \in O_l \cup S_l \\ \Delta m_r(i,j)^2 & \text{if } (i,j) \in O_r \cup S_r \\ \frac{\Delta m_l(i,j)^2 + \Delta m_r(i-d_l,j)^2}{2} & \text{if } (i,j) \in R_l \\ \frac{\Delta m_l(i+d_r,j)^2 + \Delta m_r(i,j)^2}{2} & \text{if } (i,j) \in R_r \\ 0 & \text{if } (i,j) \in T_l \cup T_r \end{cases} \tag{5.32}$$

where $\Delta m = m_v - m_{\tilde{v}}$, $v \in \{l, r\}$. Obviously, if the pixel is occluded, $sm(i, j)$ is computed only on one-side distortion. For the binocular suppression regions, the pixel degradation in one view is masked by pixels in the other view. As for the binocular rivalry regions, two different intensities in shutter way of the same 3D point are seen by viewers, and therefore the average of the pixel distortions in the two views is calculated.

In Ahmed's paper, only occluded regions and non-occluded regions were taken into account. Specifically, they used the Just Noticeable Difference (JND) model for occluded pixels, and the BJND model for the non-occluded pixels. Hence, the quality scores of the occluded pixels is calculated by

$$Q_k^{oc} = \frac{\sum_{i,j} \frac{1}{JND_k(i,j)} \times SSIM_k(i,j)}{\sum_{i,j} \frac{1}{JND_k(i,j)}}, \quad k \in \{l, r\} \tag{5.33}$$

where i and j are the pixel coordinates, $JND(i, j)_k$ and $SSIM_k(i, j)$ represent the JND value and the SSIM score of the pixel (i, j) in the k view. The overall quality score is computed as

$$Q3D^{oc} = \frac{Q_l^{oc} + Q_r^{oc}}{2} \tag{5.34}$$

5.3 Region Classification Strategy

In the same way, the quality scores of non-occluded pixels are computed by

$$Q_k^{noc}(i, j) = \frac{\frac{1}{BJND_k(i,j)} \times SSIM_k(i, j)}{\max\left(\frac{1}{BJND_k}\right)}, \quad k \in \{l, r\} \quad (5.35)$$

where $BJND_k(i,j)$ represents the BJND value of the pixel (i,j). Considering the binocular phenomena, the overall quality score is obtained by using a local information content weighting method as follows.

$$Q3D^{noc}(i, j) = e_l(i, j) \times Q_l^{noc}(i, j) + e_r(i, j - d_l(i, j)) \times Q_r^{noc}(i, j - d_l(i, j)) \quad (5.36)$$

where $e_l(i, j)$ and $e_l(i, j - d_l(i, j))$ are defined by

$$e_l(i, j) = \frac{EN_l(i, j)}{EN_l(i, j) + EN_r(i, j - d_l(i, j))} \quad (5.37)$$

$$e_r(i, j - d_l(i, j)) = \frac{EN_r(i, j - d_l(i, j))}{EN_l(i, j) + EN_r(i, j - d_l(i, j))} \quad (5.38)$$

where $EN(i, j)$ represents the local entropy at each spatial location (i, j), which is calculated within a circular window of size 11×11 moving pixel-by-pixel over the entire image. Then the perceptual quality score of the non-occluded can be obtained by averaging the value of the $Q3D^{noc}$ index map:

$$Q3D^{noc} = \frac{\sum_{i,j} Q3D^{noc}(i, j)}{N} \quad (5.39)$$

where N is the number of non-occluded pixels in the left image.

Similarly, the final 3D quality score is calculated by

$$Q3D = \alpha \cdot Q3D^{oc} + \beta \cdot Q3D^{noc} \quad (5.40)$$

where α and β are the weighting factors which are satisfied with $\alpha + \beta = 1$.

To simplify these region classification methods, some paper handled this problem only by detecting non-corresponding region (NCR) and corresponding region (CR) based on whether a pixel has correspondence in the other view. The overall NCR region of stereoscopic images consists of the NCR in the left image and right image, which is defined as follows:

$$NCR_e(x, y) = \begin{cases} 1, & \text{if the object is occluded at (x,y)} \\ 0, & \text{otherwise} \end{cases} \quad (5.41)$$

where $e \in \{L, R\}$ represents the non-corresponding region in left or right view; Exclude the above NCR areas, the rest matched regions in the estimated disparity maps are denoted as CR:

$$CR(x, y) = \begin{cases} 1, & \text{if pixel (x,y) is matched} \\ 0, & \text{otherwise} \end{cases} \quad (5.42)$$

Note that, only the original disparity is considered for this simplified region classification strategy, unlike the disparity of distorted stereopairs used in the first strategy. Then we should calculate the quality scores for the NCR and CR independently by measuring the similarities of the energy responses from the original and distorted stereoscopic images, which can be expressed by

$$Q_{NCR} = \frac{\sum_{(x,y) \in R^l_{NCR}} S^l_{NCR}(x, y) + \sum_{(x,y) \in R^r_{NCR}} S^r_{NCR}(x, y)}{N^l_{NCR} + N^r_{NCR}} \quad (5.43)$$

where $S^l_{NCR}(x, y)$ and $S^r_{NCR}(x, y)$ are the quality score for each pixel in the NCR of the left image and the right image, respectively. N^l_{NCR} and N^r_{NCR} are the number of pixels of the NCR for the left and right images, respectively.

$$Q_{CR} = \frac{\sum_{(x,y) \in R_{CR}} S_{CR}(x, y) \cdot w_e(x, y)}{\sum_{(x,y) \in R_{CR}} w_e(x, y)} \quad (5.44)$$

where $S_{CR}(x, y)$ is the quality score for each pixel in the CR of the left and right images, $w_e(x, y)$ denotes the weighting factor at location (x, y). The final index is calculated by combining Q_{CR} and Q_{NCR} into a final quality score by

$$Q = w_{CR} \cdot Q_{CR} + (1 - w_{CR}) \cdot Q_{NCR} \quad (5.45)$$

where w_{CR} is the weighting factor assigned to the CR, which can adjust the contribution of CR component to the overall quality. When $w_{CR} = 0$ means the contribution of NCR component in quality prediction is totally ignored.

However, all of these region classification strategies cannot completely conform the actual human binocular visual response. The main reasons are as follows, on the one hand, the ground truth binocular disparity cannot be obtained from the distorted stereoscopic images, which introduces some errors in region classification. On the other hand, except for binocular disparity, some other interference factors still influence the perceived depth, such as monocular depth cues and edge direction. In a word, existing technologies cannot fully simulate the binocular visual system (Mittal et al. 2011; Wang et al. 2011).

5.4 Visual Fatigue and Visual Discomfort

Taking binocular fusion and rivalry into consideration is extensive and inevitable for modern SIQA methods. However, human binocular vision response is a complex process, which contains other important binocular effects, such as binocular visual discomfort and visual fatigue. Several factors can cause visual discomfort when viewing stereoscopic images. In Tyler's paper, the issue of visual discomfort is caused by misalignment of viewed stereopairs with regard to vertical and torsional disparities (Tyler et al. 2012). In Kooi and Meesters' paper, flawed presentations of horizontal disparity, such as excessively large or otherwise unnatural disparities, can also lead to severe visual discomfort. In Lambooij's paper, various other factors can lead to the visual discomfort were discussed, including optical distortions and motion parallax. Generally, in the absence of geometrical distortions and window violations, horizontal disparity related factors mainly resulted in visual discomfort.

Visual fatigue is defined as a syndrome, whose presence is assessed by the observation of zero, one or several symptoms and zero, one or several clinical signs. Different from visual fatigue, which is supposed to have a longer rise time and a longer fall time, visual discomfort appears and disappears with a steep rise time and a steep fall time, that is to say, visual discomfort disappears rapidly when the visual task is interrupted, through asking the observer to close his eyes or stopping the visual stimulus. In this section, we will briefly introduce the assessment of visual fatigue and mainly focus on the assessment of visual discomfort.

5.4.1 Visual Fatigue Prediction for Stereoscopic Image

In general, considering visual fatigue, four main factors are taken into account, which are the geometry skewed of left and right views, excessive parallax and discontinuous change of parallax in image, poor electrical properties of stereopairs, and contradiction between convergence and accommodation, respectively. The last factor plays an essential role in the subjective aspect. As shown in the Fig. 5.4a, when a real object is observed, the sight of left and right eyes converges at the gazing point corresponding to the distance d. In Fig. 5.4b when stereoscopic videos are watched, focus adjustment is on the display screen. Therefore, the distance of accommodation is different from that of convergence, namely $d \neq D$, which makes viewers feel uncomfortable. Generally, the larger the value of $|d - D|$ is, the more uncomfortable viewers feel.

Except for the subjective factors, the objective factors mainly are closely related to the content of stereoscopic images/videos. Here we will introduce a typical objective criterion of visual fatigue, which takes spatial variation, temporal variation, and scene movement variation into consideration and then outputs an overall degree of visual fatigue, as shown in Fig. 5.5.

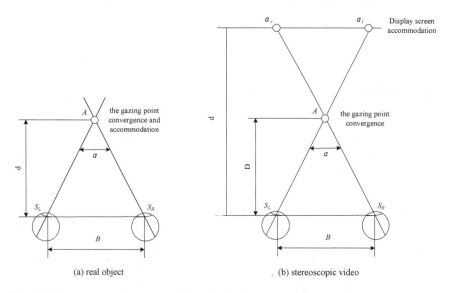

Fig. 5.4 Convergence and accommodation of real object and stereoscopic video

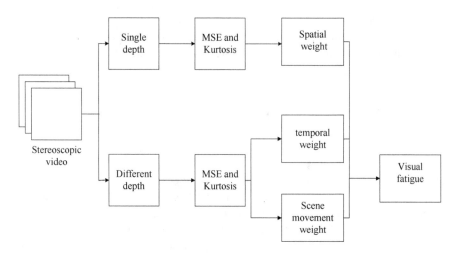

Fig. 5.5 Procedure of the objective visual fatigue assessment

In this model, spatial variation is used to represent the variation of disparity in spatial domain. Unlike traditional MSE operates on the whole image, which treats all pixels of the images equally, this model simulates the behavior of human vision system, considering different sensitivity towards different parts of the image. Specifically, it first researches the variance of 16×16 pixels in blocks of depth, as shown in Fig. 5.6. An overlapped and micromesh method is adopted to calculate MSE of

5.4 Visual Fatigue and Visual Discomfort

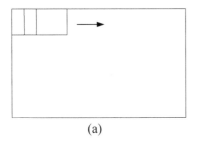

(a)

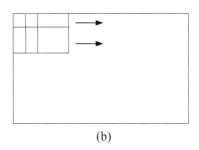

(b)

Fig. 5.6 The approach of dividing the image into blocks

each block, where the adjacent blocks share pixels at half of the block in the both horizontal and vertical directions.

In this way, MSE can be redefined as

$$MSE = \frac{1}{N*N} \sum_{x=1}^{N} \sum_{y=1}^{N} [f_i(x,y) - E(f_i(x,y))]^2 \quad (5.46)$$

where N = 16 is the size of the block, $f_i(x,y)$ is the pixel value at (x, y), $E(f_i(x,y))$ is the average pixel of each block.

Secondly, the kurtosis of each block is computed to reflex the distribution of the MSE (Caviedes and Oberti 2004), and then the weight of spatial variation can be defined as:

$$C_1 = \frac{\frac{1}{9}\sum_{i=1}^{9} k_i - k_{\min}}{k_{\max} - k_{\min}} \quad (5.47)$$

$$C_{sv} = \frac{1}{M} \sum_{j=1}^{M} C_1 \quad (5.48)$$

where k_i represents the kurtosis of ith block of depth image (Speranza et al. 2006), C_{sv} represents the average rate of change in spatial domain, M is the total number of frames of stereoscopic video. If C_{sv} is close to 0, it means the content of image is complex. While if C_{sv} is close to 1, it means the content of image is simple.

The variance of disparity difference in the temporal domain also easily leads to the visual fatigue. Thus we can calculate the disparity difference of pixels in continuous depth frames by

$$MSE_{diff} = \frac{1}{N*N} \sum_{x=1}^{N} \sum_{y=1}^{N} [diff_i(x,y) - E(diff_i(x,y))]^2 \quad (5.49)$$

where $diff_i(x, y)$ is difference value of corresponding pixels between current and previous depth images, $N = 16$, and $E(diff_i(x, y))$ is the average value of $diff_i(x, y)$. In the same calculation way as spatial variation, the temporal variation can be obtained by

$$C_2 = \frac{\frac{1}{9}\sum_{i=1}^{9} k_i - k_{min}}{k_{max} - k_{min}} \tag{5.50}$$

$$C_{TC} = \frac{1}{M-1} \sum_{j=1}^{M-1} C_2 \tag{5.51}$$

where C_{TC} represents the disparity variation in time domain, and M represents the total number of frames of the video. If the C_{TC} is close to 0, the depth frames have dynamic objects of objects. While if the C_{TC} is close to 1, the depth frames have static objects of the objects.

Unlike 2D videos, we must take camera motion into account when watching 3D videos for HVS perceiving motion more sensitively at this time. The same computing method is adopted to obtain the weight of scene movement variation as follows:

$$MSE_{diff} = E(|diff_i(x, y)|)^2 \tag{5.52}$$

$$C_3 = \frac{\frac{1}{9}\sum_{i=1}^{9} k_i - k_{min}}{k_{max} - k_{min}} \tag{5.53}$$

$$C_{SC} = \frac{1}{M-1} \sum_{j=1}^{M-1} C_3 \tag{5.54}$$

where C_{SC} represents the variation of scene movement by camera motion. If the C_{sc} is close to 1, the speed of the camera motion is low, but if the C_{sc} is close to 0, the speed of the camera motion is high.

Finally, these factors can be combined linearly to generate the overall measurement of visual fatigue as follows:

$$C = a_1 C_{SV} + a_2 C_{TC} + a_3 C_{SC} \tag{5.55}$$

where a_1, a_2, a_3 are the coefficient of linear combination.

5.4.2 Visual Discomfort Prediction for Stereoscopic Image

In many existing discomfort prediction models, most of the features are relative to disparity map, such as the disparity range, disparity distribution, maximum angular,

5.4 Visual Fatigue and Visual Discomfort

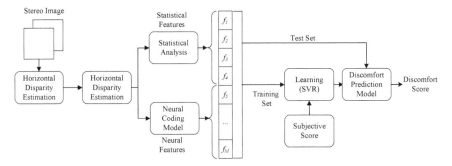

Fig. 5.7 Overall processing flow of the neural and statistical feature based 3D visual discomfort predictor

disparity gradient, disparity location, and various other quantities calculated from disparity maps (Park et al. 2014; Sohn et al. 2013; Nojiri et al. 2003; Yano et al. 2002; Choi et al. 2012; Kim and Sohn 2011). Therefore, the performance of existing discomfort assessment model is strongly dependent on the accuracy of estimated disparity map. The typical overall framework of stereoscopic image visual discomfort is shown in Fig. 5.7.

From the above figure, two types of information are calculated from the estimated horizontal map to generate a feature factor which is used to predict the degree of visual discomfort. The first type derives from a statistical analysis of horizontal map, and the second type extracts a predictive measure of neural activity in a brain center which is closely related to both horizontal disparity processing and vergence eye movement control. Then these extracted features are learned with subjective stereoscopic image discomfort scores recorded in a large human study by support vector regressor (SVR). The overall visual discomfort score is then generated by discomfort prediction model, which is trained on the IEEE standard Association stereo image database (Park et al. 2012).

After that, Ding proposed a new method of binocular visual discomfort estimation, which takes three important indices into consideration, spatial depth level, depth complexity, and experienced disparity range (Ding et al. 2019). Moreover, they make extra efforts to improve performance: (i) The observer's viewing experience is resistant to the outliers in the disparity maps, which appears in the maps as elements with extremely high or low values. So they denoted α_1 of the data as outliers and abandoned, consists of half extremely large ones and half extremely small, and the rest $(1 - \alpha_1)$ of data are retained for further processing. (ii) Hence the image regions with high saliency have a large impact on the sensation of visual discomfort and overall image quality. In the same way, α_2 of data are abandoned with relatively low saliency level, $(1 - \alpha_2)$ of data are retained for further calculation. In the next chapter we will give a detailed introduction of visual saliency. Parameters α_1 and α_2 are empirically set as 3 and 20%. Then the remaining data are reorganized as a vector d.

Let the disparity map D includes N elements and d includes n elements, where $n = (1 - \alpha_1) \cdot (1 - \alpha_2)$. The estimated visual discomfort is defined as:

$$q_3 = \left\{ d_{ave}, \frac{1}{n} \sum_i (d_i - d_{ave})^2, \frac{d_{max}}{d_{min}} \right\} \qquad (5.56)$$

where d_i denotes the ith element of d; d_{ave}, d_{max} and d_{min} represent the average, maximum and minimum of d, respectively. Finally, q_3 and other quality indices are mapped to the final stereoscopic image quality scores with a regression function trained by SVM. To summarize this kind of methods, we overview the depth estimation algorithms in the past few years in Table 5.1.

The accuracy of the horizontal disparity closely related to the performance of this kind of method, so additional information such as the range of available depths and the camera parameters are needed (Scharstein and Szeliski 2002). However, this type of information may not be available, which raises the question of whether it is possible to construct a stereoscopic visual discomfort model without disparity calculation. To solve this problem, Chen formulated a new discomfort predictive framework which used a simple, perceptually relevant implicit disparity tool, called the percentage of un-linked pixels (PUP) map (Chen et al. 2017). The detailed information about PUP map can be seen in Chen's paper, we mainly talk about how to build up visual discomfort model using PUP in this section.

Table 5.1 Overview of visual discomfort models based on disparity map

Models	Depth estimation	Extracted features
Yano et al. (2002)	Block level correlation computation	The ratio of sums of horizontal disparities near the screen and those far from the screen
Nojiri et al. (2003)	Block level phase correlation computation	The minimum and maximum values, range, dispersion, absolute average and average disparities
Kim and Sohn (2011)	SIFT matching and region-dividing technique with energy-based regularization	The experienced horizontal disparity range and maximum angular disparity
Choi et al. (2012)	Optical flow software	Anomaly of AV/A ratio, anomaly of VA/V ratio, absence of de-focus blur, and absence of differential blur
Park et al. (2014)	Dynamic programming	Spatial depth complexity and depth position
Park et al. (2015)	Estimated horizontal disparity map	Statistical and neural features
Ding et al. (2019)	Exclude outliers and use visual saliency	Spatial depth level, depth complexity, and experienced disparity

5.4 Visual Fatigue and Visual Discomfort

The basic principle of this model is that a corresponding image pair on retinal having more pixels with different orientations, luminances, and other image properties will have more un-linked pixels and a high PUP (Howard and Rogers 2012; Sperling 1970). Then we classify pixels into groups where the pixels in the same group have similar features. These groups are defined as feature groups. Specifically, the number of un-linked pixels can be computed by counting the number of pixels in different feature groups:

$$S_{unlinked} = \frac{\sum_{i=1}^{N_{hist}} \left| H^{l,N_{hist}}(i) - H^{r,N_{hist}(i)} \right|}{2} \quad (5.57)$$

where N_{hist} is the number of feature groups, $H^{l,N_{hist}}(i)$ and $H^{r,N_{hist}}(i)$ are the number of pixels in the ith feature group of the left and right views. Thus PUP is defined as:

$$PUP^{N_{hist}} = \frac{\sum_{i=1}^{N_{hist}} \left| H^{l,N_{hist}}(i) - H^{r,N_{hist}(i)} \right|}{2N_{total}} \quad (5.58)$$

In this method, the width of the image patch should be set reasonably to capture large disparities. That's because images with highly variable disparity levels and rich details will be better analyzed by adopting a smaller patch width, while images with few disparity levels and details will be better analyzed by adopting a larger width. Hence, we need multiple sizes of patch widths for accurate modeling. Generally, three different patches are adopted in this model. The width of the largest patch corresponds to the disparity limit of the fusional area (Yeh and Silverstein 1990). The width of the smallest patch corresponds to the disparity limit of the comfort zone (Percival 1892). The width of the average patch corresponds the average value of the fusional limit and the comfort limit disparities. Specifically, for an image with a resolution of 1920 × 1080 pixels, the corresponding block widths of PUP-L, PUP-M, PUP-S are calculated to be 280, 168 and 56 pixels respectively. Finally, to predict the degree of visual discomfort of stereoscopic image, each of four kinds of features are extracted from the PUP-L, PUP-A and PUP-S maps respectively with the feature extraction method in Park et al. (2014). Therefore, considering three different PUP maps, the total number of features used in prediction is 12. Four kinds of features contain the average values of positive and negative PUP, and the upper and lower 5% of the PUP values, which are defined as follows:

$$f_1 = \frac{1}{N_{Pos}} \sum_{PUP(n)>0} PUP(n) \quad (5.59)$$

$$f_2 = \frac{1}{N_{Neg}} \sum_{PUP(n) \leq 0} PUP(n) \quad (5.60)$$

$$f_1 = \frac{1}{N_{5\%}} \sum_{n \leq N_{\text{total}} \times 0.05} PUP(n) \qquad (5.61)$$

$$f_1 = \frac{1}{N_{95\%}} \sum_{n \geq N_{\text{total}} \times 0.95} PUP(n) \qquad (5.62)$$

where $PUP(n)$ is the nth smallest value in a PUP map, N_{Pos} and N_{Neg} are the total number of positive and negative values in a PUP map.

Except for PUP based visual discomfort model, Ding and Zhao also propose a scheme that operates in a patch-based manner using adaptive image segmentation without disparity map. The adaptive segmentation is an iterative process. Let each image pixel as a five-dimensional point, which are two space dimensions represent the location of the pixel and three-color dimensions represent Red, Green, and Blue channels respectively. Supposed there are N pixels in an image, which are divided into K groups through the K-means algorithm. The special steps of this algorithm are as follows.

Firstly, K points are randomly selected from N points to serve as clustering center. Then the following two steps are iteratively performed until the clustering centers stop moving or the iteration time is over maximum (here set as 1000):

Step one: Assign each point to the cluster which is closest to it. The cost function J can be minimized by assigning points to different clusters.

$$J = \sum_k \sum_i r_{i,k} \| t_i - c_k \|^2 \qquad (5.63)$$

where $r_{i,k}$ is a judgment function which equals 1 if point t_i belongs to the kth cluster, and 0 otherwise, c_k is the kth clustering center.

Step two: After all points are assigned to certain clusters, the clustering centers are updated to the geometrical center of the points that belongs to the cluster:

$$c_k = \frac{1}{N_k} \sum_i r_{i,k} \cdot t_i \qquad (5.64)$$

where N_k is the number of points belonging to cluster k.

The parameter K in this algorithm has great impact on the segmenting results. Specifically, we choose four certain values for K to compose a numeric sequence with common ratio α. The smaller K is usually adopted in higher scales, for the scenes are displayed in a more macro manner. The numbers of segments K_s are defined as

$$K_s = \left\lfloor \frac{K_1}{\alpha^{s-1}} \right\rfloor \qquad (5.65)$$

5.4 Visual Fatigue and Visual Discomfort

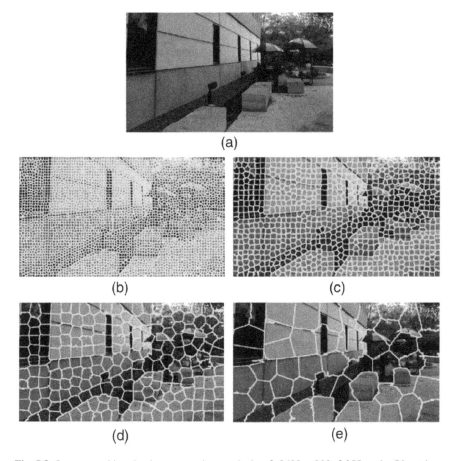

Fig. 5.8 Image **a** and its adaptive segmenting results into **b** 3600, **c** 900, **d** 255, and **e** 56 patches

where $s = 1, 2, 3, 4$, respectively denote different scales of K; K_1 and α are set as 3600 and 4, respectively; We can see all the four scales of a monocular view and its segmenting results in Fig. 5.8.

From the above figures, we can find that similar pixels in a neighborhood are grouped into patches. An object appears on the left and right retinas always accompanied by a horizontal position shift, thus the corresponding segmented patches naturally obey the same rules. So we can use the position shifts of corresponding patches to reflect the size of binocular disparity and therefore an approximation for the degrees of visual discomfort.

The specific process is operated on each row independently, for the position shifts of objects are limited in horizontal direction. The pixels of each row are labeled different numbers according to the cluster they belong to. For example, we define the pixel at the left-most as 1, and gradually increase the number by 1 each time when a new cluster meets. Suppose a monocular view contains X rows and Y columns, the total pixels N can be computed by

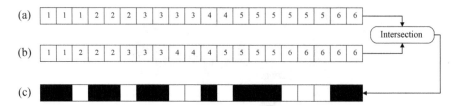

Fig. 5.9 Illustration of two vectors (**a**) and (**b**) generating their intersection (**c**)

$$N = X \cdot Y \tag{5.66}$$

Then each row of the monocular view can be labeled as a Y-size vector, denoting as $v_{L,x}$ and $v_{R,x}$ for the left view and right view respectively. VD_x is the contribution of this row to visual discomfort, which can be calculated by

$$VD_x = \frac{1}{Y} \sum_y \text{sgn}(v_{L,x,y}, v_{R,x,y}) \tag{5.67}$$

where y denotes the yth element of a vector, signal function *sgn* outputs 1 when its two inputs have equal values, and 0 otherwise.

To more intuitively display how the intersection function works between vectors of left view and right view, a straightforward figure is shown in Fig. 5.9. Vector (a) and vector (b) represent the vector from left and right views, and the position shift between (a) and (b) is noted in (c).

As for the entire image, the overall visual comfort denoted as q^d is estimated by averaging the discomfort contributions of all rows, which is computed by

$$q^d = \frac{1}{X} \sum_x VD_x \tag{5.68}$$

Compared to single scale results, the synthesis of results from four scales will give a more comprehensive description of position shifts and therefore a more accurate prediction of visual discomfort is obtained.

5.5 Summary

Compared with traditional 2D videos, the emergence of 3D videos has brought about new technical challenges, especially in terms of good end-user 3D quality of experience (QoE) (Mittal et al. 2011). Obviously, consumers have high expectations for the QoE closely related with the services they receive. In recent years, QoE evaluation has played an important role in SIQA techniques. However, the methods discussed in the last chapter only considered the different weighting factors of the left and right view, and totally ignored viewer's 3D QoE factors. The degree of image quality obtained in those methods could not accurately reflect human subjective evaluation results. To better utilize 3D perceptual properties in the human visual system, Maalouf and Larabi (2011) first generated a "Cyclopean" image from left and right views to model the perception by the HVS, rather than directly applied 2D IQA on left and right images respectively. And then the researchers focused on building appropriate models of this intermediate image. The most typical example was proposed by Chen, which took binocular rivalry into account using Levelt's linear model. After that, many approaches have been improved on the basis of Chen's method, such as replacing Gabor filter with Log-Gabor filter, employing different stereo algorithms, adopting different linear combination models. Except for these methods, some algorithms simultaneously used cyclopean image, the left image and right image to generate overall subjective quality scores. Some methods extracted features from distorted and reference images and then only use these features to build prediction model. And some methods even developed blind quality estimator for stereo images based on binocular combination and extreme learning machine.

However, a detail should attract out attention: a pixel may have no matched pixel in the corresponding region of the other view, and monocular vision will occur in this non-corresponding area. To more accurately simulate human perception process, some methods adopted region classification strategy. Specifically, the distorted 3D image were categorized into non-corresponding region, binocular fusion region and binocular suppression region, or the latter two were merged into corresponding region.

Finally, visual fatigue and visual discomfort will also affect people's judgment of image quality. We have discussed the main factors that cause visual fatigue and visual discomfort, and the accuracy of estimated disparity map determine the performance of visual fatigue or discomfort model. Actually, the improved accuracy of disparity calculation will increase calculation complexity and need additional information. To overcome these limitations, some researches designed PUP maps or adaptive image segmentation-based methods as alternative, and the speed and performance of these algorithms are improved.

References

Ahmed FS, Larabi M-C, Mohamed FK (2014) Stereoscopic image quality metric based on local entropy and binocular just noticeable difference. In: International conference on image processing, Paris, France, pp 2002–2006

Anderson PA, Movshon JA (1989) Binocular combination of contrast signals. Vision Res 29:1111–1132

Blake R, Wilson H (2011) Binocular vision. Vision Res 51(7):754–770

Bolanowski SJ (1987) Contourless stimuli produce binocular summation. Vision Res 27:1943–1951

Bourassa CM, Rule SJ (1994) Binocular brightness: a suppression-summation trade off. Can J Exp Psychol 48:418–433

Caviedes J, Oberti F (2004) A new sharpness metric based on local kurtosis, edge and energy information. Sig Process Image Commun 19(2):147–161

Chen MJ, Su CC, Kwon DK, Kwon DK, Cormack LK et al (2013) Full-reference quality assessment of stereopairs accounting for rivalry. Signal Process Image Commun 28(9):1143–1155

Chen J, Zhou J, Sun J, Bovik AC (2017) Visual discomfort prediction on stereoscopic 3D images without explicit disparities. Sig Process Image Commun 51:50–60

Choi J, Kim D, Choi S, Sohn K (2012) Visual fatigue modeling and analysis for stereoscopic video. Opt Eng 51(1):017206

Cogan AI (1982) Monocular sensitivity during binocular viewing. Vision Res 22:1–16

Cogan AI (1987) Human binocular interaction: towards a neural model. Vision Res 27:2125–2139

Curtis DW, Rule SJ (1978) Binocular processing of brightness information: a vector-sum model. J Exp Psychol Hum Percept Perform 4(1):132–143

Ding J, Sperling G (2006) A gain-control theory of binocular combination. Proc Natl Acad Sci U S A 103(4):1141–1146

Ding Y, Zhao Y, Chen X, Zhu X, Andrey K (2019) Stereoscopic image quality assessment by analysing visual hierarchical structures and binocular effects. IET Image Proc 13(10):1608–1615

Engel GR (1967) The visual processes underlying binocular brightness summation. Vision Res 7(9):753–767

Field DJ (1987) Relations between the statistics of natural images and the response properties of cortical-cells. J Opt Soc Am A Opt Image Sci Vision 4(12):2379–2394

Fleet DJ, Wagner H, Heeger DJ (1996) Neural encoding of binocular disparity: energy models, position shifts and phase shifts. Vision Res 36(12):1839–1857

Gorley P, Holliman N (2008) Stereoscopic image quality metrics and compression. Electron Imaging 6803:680305

Grossberg S (1987) Cortical dynamics of three-dimensional form, color and brightness perception: II. Binocular theory. Percept Psychophys 41:117–158

Grossberg S (1994) 3-D vision and figure-ground separation by visual cortex. Percept Psychophys 55:48–120

Grossberg S (1997) Cortical dynamics of three-dimensional figure-ground perception of two-dimensional pictures. Psychol Rev 104:618–658

Hachicha W, Beghdadi A, Cheikh FA (2012) Combining depth information and local edge detection for stereo image enhancement. In: European signal processing conference, Bucharest, Romania, pp 250–254

Hachicha W, Beghdadi A, Cheikh AF (2013) Stereo image quality assessment using a binocular just noticeable difference model. In: International conference on image processing, Melbourne, Australia, pp 113–117

Henkel RD (1997) Fast stereovision by coherence detection. In: Lecture notes in computer science, pp 297–304

Howard IP, Rogers BJ (2008) Seeing in depth. Oxford University Press, New York, USA

Howard IP, Rogers BJ (2012) Perceiving in depth. Stereoscopic Vis 2:1–656

Julesz B (1971) Foundations of cyclopean perception. University Chicago Press, Chicago, IL, USA

Kim D, Sohn K (2011) Visual fatigue prediction for stereoscopic image. IEEE Trans Circuits Syst Video Technol 21(2):231–236

Klaus A, Sormann M, Karner K (2006) Segment-based stereo matching using belief propagation and a self-adapting dissimilarity measure. In: International conference on pattern recognition, Long Quan, Hong Kong, pp 15–18

Kolmogorov V, Zabih R (2001) Computing visual correspondence with occlusions using graph cuts. In: International conference on computer vision, pp 508–515

Lehky SR (1983) A model of binocular brightness and binaural loudness perception in humans with general applications to nonlinear summation of sensory inputs. Biol Cybern 49:89–97

Levelt WJM (1968) On binocular rivalry. Mouton, The Hague, Paris

References

Li K, Shao F, Jiang G, Yu M (2014) Full-reference quality assessment of stereoscopic images by learning sparse monocular and binocular features. In: Proceedings of the international society for optical engineering, pp 927312–1927312-10

Lin YH, Wu JL (2014) Quality assessment of stereoscopic 3D image compression by binocular integration behaviors. IEEE Trans Image Process 23:1527–1542

Lin YC, Yang JC, Lu W, Meng QG, Lv ZH et al (2017) Quality index for stereoscopic images by jointly evaluating cyclopean amplitude and cyclopean phase. IEEE J Sel Top Sig Process 11(89):101

Maalouf A, Larabi MC (2011) CYCLOP: a stereo color image quality assessment metric. In: Proceedings of IEEE international conference on acoustics, speech, and signal processing, Prague, Czech Republic, pp 1161–1164

MacLeod DIA (1972) The Schrodinger equation in binocular brightness combination. Perception 1:321–324

Mather G (2008) Foundations of sensation and perception. Psychology Press, Oxon, UK

Mittal A, Moorthy AK, Ghosh J, Bovik AC (2011) Algorithmic assessment of 3D quality of experience for images and videos. In: Proceedings of IEEE digital signal process, pp 338–343

Nojiri Y, Yamanoue H, Hanazato A, Okano F (2003) Measurement of parallax distribution and its application to the analysis of visual comfort for stereoscopic HDTV. Electron Imaging 195–205

Ojala T, Pietikäinen M, Mäenpää T (2002) Multiresolution gray-scale and rotation invariant texture classification with local binary patterns. IEEE Trans Pattern Anal Mach Intell 24(7):971987

Park J, Oh H, Lee S (2012) IEEE-SA stereo image database [Online]. Available: http://grouper.ieee.org/groups/3dhf/

Park J, Lee S, Bovik AC (2014) 3D visual discomfort prediction: vergence, foveation, and the physiological optics of accommodation. IEEE J Sel Top Sig Process 8(3):415–427

Park J, Oh H, Lee S, Bovik AC (2015) 3D visual discomfort predictor: analysis of disparity and neural activity statistics. IEEE Trans Image 24 (3):1101–1114. https://doi.org/10.1109/TIP.2014.2383327

Percival AS (1892) The relation of convergence to accommodation and its practical bearing. Ophthal Rev 11:313–328

Scharstein D, Szeliski R (2002) A taxonomy and evaluation of dense two-frame stereo correspondence algorithms. Int J Comput Vision 47(1):7–42

Schrodinger E (1926) Die gesichtsempfindungen. Mueller-Pouillets Lehrbuch der Physik 2(1):456–560

Shao F, Li K, Lin W, Jiang G (2015) Full-reference quality assessment of stereoscopic images by learning binocular receptive field properties. IEEE Trans Image Process 24(10):2971–2983

Shao F, Chen W, Jiang G, Ho Y-S (2017) Modeling the perceptual quality of stereoscopic images in the primary visual cortex. IEEE Access 5:15706–15716

Sohn H, Jung YJ, Lee S-I, Ro YM (2013) Predicting visual discomfort using object size and disparity information in stereoscopic images. IEEE Trans Broadcast 59(1):28–37

Speranza F, Tam WJ, Renaud R, Hur N (2006) Effect of disparity and motion on visual comfort of stereoscopic images. In: Proceedings of SPIE-IS&T electronic imaging, p 6055

Sperling G (1970) Binocular vision: a physical and a neural theory. Am J Psychol 461–534

Steinman S, Steinman B, Garzia R (2000) Foundations of binocular vision. A clinical perspective. McGraw-Hill, New York, USA

Sun D, Roth S, Black MJ (2010) Secrets of optical flow estimation and their principles. In: Conference on computer vision and pattern recognition, pp 2432–2439

Sun G, Ding Y, Deng R, Zhao Y, Chen X, Krylov SA (2018) Stereoscopic image quality assessment by considering binocular visual mechanisms. IEEE Access 6:511337–511347

Tyler CW, Likova LT, Atanassov K, Ramachandra V, Goma S (2012) 3D discomfort from vertical and torsional disparities in natural images. In: Proceedings of society of photo-optical instrumentation engineers, vol 8291, pp 82910Q-1–82910Q-9

Wang X, Kwong S, Zhang Y (2011) Considering binocular spatial sensitivity in stereoscopic image quality assessment. Visual Commun Image Process 1–4

Xu HY, Yu M, Luo T, Zhang Y, Jiang GY (2016) Parts-based stereoscopic image quality assessment by learning binocular manifold color visual properties. J Electron Imaging 25:061611

Yano S, Ide S, Mitsuhashi T, Thwaites H (2002) A study of visual fatigue and visual comfort for 3D HDTV/HDTV images. Displays 23(4):191–201

Yasakethu SLP, Hewage CTER, Fernando WAC, Kondoz AM (2008) Quality analysis for 3D video using 2D video quality models. IEEE Trans Consum Electron 54(4):1969–1976

Yeh Y-Y, Silverstein LD (1990) Limits of fusion and depth judgment in stereoscopic color displays. Factors Ergon 32(1):45–60

Zhang Y, Chandler DM (2015) 3D-MAD: a full reference stereoscopic image quality estimator based on binocular lightness and contrast perception. IEEE Trans Image Process 24:3810–3825

Zhao Y, Chen Z, Zhu C, Tan YP, Yu L (2011) Binocular JND model for stereoscopic images. IEEE Sig Process 18(1):19–22

Zhou W, Yu L, Zhou Y, Qiu W, Wu M, Luo T (2017) Blind quality estimator for 3D images based on binocular combination and extreme learning machine. Pattern Recogn 71:207–217

Chapter 6
Stereoscopic Image Quality Assessment Based on Human Visual System Properties

Abstract Modelling the behavior of Human Visual System (HVS) is the ultimate target of Image Quality Assessment (IQA). The hierarchical structure of HVS and different HVS models are introduced firstly in this chapter. And some classical IQA methods based on the hierarchical structure of HVS are discussed in detail. Visual attention, as one of the most important mechanisms of the HVS, is clarified clearly and some Stereoscopic Image Quality Assessments (SIQA) methods based on visual saliency are also presented. In the end of this chapter, Just Noticeable Difference (JND) model and corresponding IQA methods are introduced.

Keywords Stereoscopic image quality assessment · Human visual system · Visual saliency · Just noticeable difference

6.1 Introduction

Since the human visual system (HVS) is the ultimate receiver of image information, assessing the perceptual quality of nature images, to a large extent, depends on the relevant degree between the designed image quality assessment (IQA) algorithm and HVS. Modelling the behavior of HVS is the target of IQA fields, as well in many research fields in image processing, computer vision and artificial intelligence. In the past decades, advances in human vision have been derived by a series of psychophysical and physiological research, which improve our understanding of the mechanisms of the HVS and enable us to express these psychophysical findings as mathematical models. Motivated by this, in IQA for plane images, researchers have designed a number of methods to predict the perceptual quality of the image mainly by exploring and analyzing the various functional aspects of the HVS. For example, some well-established IQA models that address the lower level aspects of early vision (e.g., contrast sensitivity, luminance masking, and texture masking) are claimed to be much more reliable and accurate than the purely mathematical and statistical models, such as peak signal-to-noise ratio (PSNR) and structure similarity (SSIM). By utilizing the observation that the HVS is very suitable for extracting the structural information from visual scenes, the subsequent experiments of IQA have demonstrated that these HVS-based methods can achieve higher consistency with subjective judgements than before. However, the mechanisms of HVS are still too mysterious to be imitated by several simple functional operations, even if significant progress has been made in psychophysical research. It is safe to state that the research of IQA and stereoscopic image quality assessment (SIQA) could benefit significantly by applying the biological characteristics of HVS, although there remains limited in modelling of HVS due to lack of knowledge of the HVS.

So far, the biological responses of the HVS maintain a mystery for researchers. Therefore, the physiological research on the HVS remains studying. Fortunately, based on continuous in-depth research on the human visual system, researchers have gradually explored some characteristics of HVS, among which the hierarchical structure of HVS is recognized as one of the most important physiological discoveries in vision research in the past few decades (Hubel and Wiesel 1968). Section 6.2 introduced the hierarchical structure of HVS in detail and meanwhile different hierarchical models of HVS were built by the computer vision community for effectively simulating the characteristics of HVS. Concretely, the HVS consists of several certain areas of visual cortex, each of which has a different visual processing function. Through the simulation and synthesis of different visual areas, we can finally obtain the comprehensive visual results (Kandel et al. 2000). Research has shown that the hierarchical solution is very helpful for SIQA. So we will introduce some classical SIQA methods that draw inspiration from the hierarchical structure of HVS in Sect. 6.3.

As one of the most important visual properties in HVS, the mechanism of visual attention is investigated for improving the reliability of IQA methods in current research, which refers to the visual process that HVS could select the most attractive visual information from natural images. Given a visual sense, human eyes are more likely to be attracted by the center area or the area with high contrast, which is named visual saliency. Research on IQA based on visual saliency has been explored for a long time, although the fact that how visual saliency affects image quality perception is unknown. It is certain that visual saliency has a great influence on image perceptual quality. Specifically, the effect of visual salience, reflected in the task of IQA in general, can be characterized by the assumption that distortion occurs in salient area is more annoying than any other areas.

The last visual property we would like to discuss is just noticeable difference (JND), which is motivated by the visual masking properties of HVS. JND is a significant property of HVS reflecting the maximum amount of distortion that the HVS can perceive. Given a distortion image, only the distorted changes exceed the visibility threshold could be recognized by human eye, leading to the quality degradation of images (Legras et al. 2004). Thus, the property of JND plays an important role in characterizing the perceptual quality of distorted images, which is very suitable for the tasks of IQA. There has been many research incorporating JND characteristic and other important HVS properties for improving its performance over the past few years. The detailed introduction about JND will be given in Sect. 6.4.

Finally, Sect. 6.5 concludes summary and related discussions about HVS-based SIQA.

6.2 Human Visual System

According to the psychophysical and physiological research, biological visual systems, in particular HVS, can accomplish the tasks of computer vision almost effortlessly, since biological visual systems are the ultimate receiver of nature scenes. Therefore, the HVS has often been utilized as an inspiration for the computer vision researchers, as well as for the task of SIQA. The HVS is so complicated that the researchers are lacking in more detailed understanding of the specific anatomy structure and the process stage of visual signals in the HVS, becoming an insurmountable obstacle to further progress in SIQA. Due to the importance of HVS to computer vision tasks, a large number of researchers have explored human visual system in the past few years, revealing the essence of the HVS to the greatest extent. (Hubel and Wiesel 1962) built the analogous hierarchical structure of biological visual system by exploring the mechanism of cat's visual generation. The authors found that cats could form a cognition of the whole nature scene by extracting different visual information in different visual levels. Based on the subsequent works on (Marr 1983; Tenenbaum et al. 2001; Bengio 2009), the computer vision community began to build the hierarchical models of HVS and simulated the biological levels of HVS.

The simplified anatomy structure of HVS and corresponding flow of neural processing of visual information are shown in Fig. 6.1. The processing stages of visual signals can roughly consist of four stages: optical processing, retina, lateral geniculate nucleus (LGN) and visual cortex. As is shown in Fig. 6.1a, the HVS begins

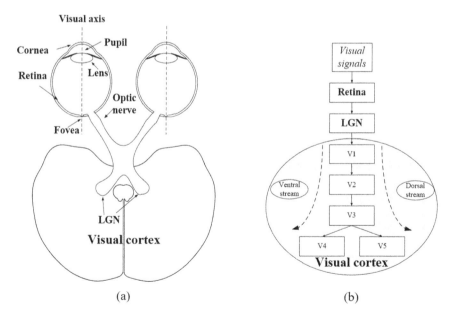

Fig. 6.1 **a** simplified anatomy structure of HVS and **b** corresponding simplified signal transmitting flow in brain

from human eyes, and goes through the refractive system for optical processing, including the cornea, the pupil and the lens. Finally, the optical visual information captured from the real-world including lightness and chrominance is perceived by the cone and rod cells in the retina, and transformed into nerve stimuli. The nerve stimuli is then transmitted to LGN, which likes a "bridge" serving the retinal and visual cortex. LGN contains two types of cells: M-type geniculate cells and P-type geniculate cells, which respond to the movement and spatial information, respectively. After the visual processing stages of LGN, the visual signals are transmitted to the visual cortex of the brain. The visual cortex of the brain performs advanced visual processing such as visual understanding and analysis by the received signals, which is the most complicated part in the HVS.

So far, the research of visual cortex remains studying and exploring. Fortunately, it has been proved that visual cortex has a hierarchical structure, consisting of the primary visual (V1) area, the secondary visual (V2) area, V3, V4, etc. The V1 area is the basis of early visual perception, occupying the largest area of the visual cortex. The V1 area can deal with the low-level visual features such as edge, spatiotemporal frequency, disparity and so on, mainly responsible for visual perception of the electrical signals transmitted from the LGN. In addition, in V1 area, two types of cells are included: simple cells and complex cells, in which the former is sensitive to phase and position information while the latter has a larger spatial receptive field than the former (Orban 2008). As the visual area occupying the second largest area of the visual cortex, the V2 area follows the V1 area and extracts high-level visual features such as texture and contour from the visual signals received from V1 area by point-to-point reception. The area composed of the V1 and V2 areas occupies 60% of the whole visual cortex, mainly forming early visual perception. The visual features perceived from the V1 and V2 areas will be further transmitted to the high-level visual areas such as V3, V4 and V5 areas for high-level visual perception.

For human visual perception, there are roughly two visual pathways for visual information processing in neural cortex, called dorsal stream and ventral stream respectively (Krüger et al. 2013). As can be seen from Fig. 6.1b, the dorsal stream starts from V1 area, goes through V2 and V3 areas, and finally arrives at V5 area, which is sensitive to action information. In contrast, the ventral stream contains V1, V2, V3 and V4 areas, representing for the perception and recognition of natural scenes. Visual cortex is too complex for researchers to have a particularly deep understanding of the visual cortex. So far, people only have a deeper understanding of the V1 and V2 areas that formed early visual perception, while the research on high-level visual perception systems is still in its infancy. Fortunately, the researchers generally believe that SIQA focus on the early vision and does not involve the advanced visual systems. Therefore, the studies on visual cortex concentrate on the V1 and V2 areas in SIQA tasks.

In addition to the anatomy structure and the physiological properties of HVS introduced above, the psychological factors of humans will also affect the results of visual perception. Therefore, the psychological research of HVS needs to be expanded to enhance the understanding of HVS. So far, the researchers have explored some psychological characteristics of HVS and established related mathematical models to represent the relationship between the psychological phenomena and visual perception, such as multi-scale energy responses, visual saliency and just-noticeable-difference. These psychological properties will be introduced in the following sections in detail. Meanwhile, for better understanding the effect of the psychological properties on SIQA, some classical SIQA methods will also be introduced with respect to the psychological models.

6.3 SIQA Based on Hierarchical Structure

In Sect. 6.2, we have known that HVS can be regarded as a hierarchical structure. Therefore, an example we would like to introduce is (Ding and Zhao 2018) to model the hierarchical structure of HVS, which has been discussed in last Chapter. Besides the visual discomfort, we mainly focus on the application of hierarchical structure of HVS here. As is known to all, V1 area is sensitive to the low-level features (e.g., edge, bars and spatial frequency information), while V2 area is more likely to be attracted by the high-level visual features such as textures, contours, shapes and so on. Therefore, the authors extracted a series of visual quality-aware features according to the mechanism of V1 and V2 areas in the human brain, including spatial frequency information and texture features.

According to the research of visual cortex (Jones and Palmer 1987), in the V1 area, each eye's receptive fields (RFs) of simple cells can be modelled by Gabor filter. Meanwhile, the response of complex cells is also proved to be similar to the magnitude of the quadrature pair of Gabor. Therefore, Gabor filter can be used for modelling the energy response of cells in V1 area, which is expressed as:

$$G(x, y; f, \theta) = \frac{f^2}{\pi \gamma \eta} \exp\left(-\left(\frac{f^2}{\gamma^2}x'^2 + \frac{f^2}{\eta^2}y'^2\right) + j2\pi f x'\right) \quad (6.1)$$

where f and θ are the RF's frequency and orientation, γ and η denote the parameters for the Gaussian function, representing the bandwidth of the filter. x' and y' are the pixels coordinates with the definition described by:

$$x' = x \cos \theta + y \sin \theta, \quad y' = -x \sin \theta + y \cos \theta \quad (6.2)$$

Other filters, such as difference of Gaussian (DOG) and log-Gabor filters also can approximate the RFs of HVS. The DOG model can be formulated as:

$$G(i, j; \sigma) - G(i, j; k\sigma) \quad (6.3)$$

with

$$G(i, j; \sigma) = \frac{1}{\sqrt{2\pi\sigma^2}} \exp\left(-\frac{i^2 + j^2}{2\sigma^2}\right) \quad (6.4)$$

where σ denotes the standard deviation of Gaussian filter, and k is the space constant that is generally equal to being 1.6.

Log-Gabor filter (Field 1987), as an improved version of Gabor filter, can make up for the disadvantages of Gabor filter including the limitation on bandwidth and non-zero component. Log-Gabor filter is usually recognized as an effective tool in extracting spatial frequency information from an image, which can be represented as:

$$LG(f, \theta) = \exp\left(-\frac{(\log(f/f_0))^2}{2(\log(\sigma_f/f_0))^2} - \frac{(\theta - \theta_0)^2}{2\sigma_\theta^2}\right) \quad (6.5)$$

where f_0 and θ_0 are the central frequency and orientation angle, and the parameters σ_f and σ_θ denote the filter's scale and angular bandwidths, respectively. The variables f and θ define the spatial position in the polar coordinates. Log-Gabor filter can be utilized to construct feature maps of spatial frequency information. For a more illustration of feature extraction, an example using log-Gabor filter is given in Fig. 6.2.

Local image textures can be represented by local binary pattern (LBP), which performs well in pattern and texture recognition (Ojala et al. 2002). By comparing the values between the central pixel and its 8 local neighbor pixels and concatenating the results into an 8-bit binary code, the value of LBP is obtained:

$$C(p_c) = \sum_{i=0}^{7} \text{sgn}(p_c, p_i) \times 2^i \quad (6.6)$$

where p_i denotes the i-th neighboring pixel of the center pixel pc, and the function sgn() is defined as:

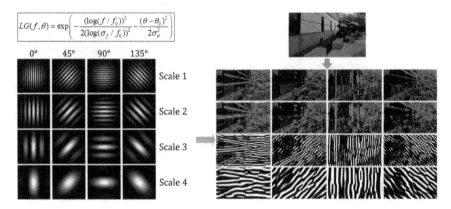

Fig. 6.2 The example of log-Gabor filter bank and its responses of an image with four scales and four orientations

6.3 SIQA Based on Hierarchical Structure

$$\text{sgn}(x, y) = \begin{cases} 1, & x > y \\ 0, & x \leq y \end{cases} \tag{6.7}$$

For the central pixel with 8 neighbor pixels, the LBP can produce highly 256 dimensional results. To reduce the dimension of the output and meanwhile achieve the rotation invariance, a rotation invariance LBP is designed:

$$LBP_{RI} = \min\{ROR(LBP, i) | i = 0, 1, \ldots, 7\} \tag{6.8}$$

where $ROR(x, i)$ defines the circular bit-wise right shift operating on the number x with the i-bits shift.

A further improved version of LBP is uniform invariance LBP. The definition of 'uniform' is based on the fact that over 90% of the 3 × 3 LBPs contain no more than two inversions in its corresponding 0-1 binary code. Therefore, these "non-uniform' LBPs (e.g., the LBP with '01001101' and '10110010') can be grouped into a category. The procedure of classifying the uniform and non-uniform 8 bits LBPs can be described as:

$$LBP_{URI} = \begin{cases} \sum_{p=0}^{7} \text{sgn}(g_p, g_c), & \text{if } U(LBP_{RI}) \leq 2 \\ 8, & \text{otherwise} \end{cases} \tag{6.9}$$

where $U(x)$ is an operation to determine the internal inversions of a binary code.

By applying log-Gabor filter and uniform rotation invariance LBPs, multi-scale frequency magnitude maps and texture maps were generated from the left and right views of a stereopair. In addition, the features representing the effects of visual discomfort were also considered and extracted. Then, these quality-aware feature maps were concatenated into cyclopean forms and mapped into the final global score using support vector regression (SVR).

In another example proposed in (Gu et al. 2019), various low-level and high-level features were considered for visual quality representations, including edge information and texture analysis.

Sobel operator is one of the most common used edge detection operators in image processing fields.

$$S_h = \begin{bmatrix} -1 & 0 & 1 \\ -2 & 0 & 2 \\ -1 & 0 & 1 \end{bmatrix}, \quad S_v = \begin{bmatrix} -1 & -2 & -1 \\ 0 & 0 & 0 \\ 1 & 2 & 1 \end{bmatrix} \tag{6.10}$$

By convolving the targeted image with Sobel operator, the horizontal and vertical gradient maps G_h, G_v can be obtained:

$$G_h = I \otimes S_h, \quad G_v = I \otimes S_v \tag{6.11}$$

where $\otimes$ represents the convolution procedure.

To better represent the visual contents of an image, a common strategy is to combine the horizontal and vertical gradient maps as the gradient magnitude map G_M and gradient orientation map G_O,

$$G_M(i,j) = \sqrt{G_h^2(i,j) + G_v^2(i,j)}$$
$$G_O(i,j) = \arctan \frac{G_v(i,j)}{G_h(i,j)} \quad (6.12)$$

The classical patterns for texture analysis such as LBPs and its improved versions uniform and rotation invariance LBPs have mentioned before. Further, the authors adopted the novel second-order local tetra pattern (LTrP) (Murala et al. 2012) to extract texture information of the targeted image in four directions by utilizing $0°$ and $90°$ derivatives of local derivative patterns (LDPs). To illustrate the second-order LTrPs, the first-order $0°$ and $90°$ derivatives must be introduced first and the definitions of them are written in:

$$I_{0°}^1(p_c) = I(p_h) - I(p_c), \quad I_{90°}^1(p_c) = I(p_v) - I(p_c) \quad (6.13)$$

where p_h and p_v are the horizonal and vertical neighboring pixels of the central pixel p_c, respectively.

Then, four directions of the p_c can be defined as:

$$I_{Dir}^1(p_c) = \begin{cases} 1, & I_{0°}^1(p_c) \geq 0 \text{ and } I_{90°}^1(p_c) \geq 0 \\ 2, & I_{0°}^1(p_c) < 0 \text{ and } I_{90°}^1(p_c) \geq 0 \\ 3, & I_{0°}^1(p_c) < 0 \text{ and } I_{90°}^1(p_c) < 0 \\ 4, & I_{0°}^1(p_c) \geq 0 \text{ and } I_{90°}^1(p_c) < 0 \end{cases} \quad (6.14)$$

According to the different directions of each central pixel, four direction maps can be obtained by:

$$DIR_i(p_c) = \begin{cases} 1, & \text{if } I_{Dir}^1(p_c) = i \\ 0, & \text{otherwise} \end{cases}, \quad i = 1, 2, 3, 4 \quad (6.15)$$

The authors have extracted four direction maps. However, this cannot reflect the magnitude information from each direction map. The fact has been proved that combing direction information and magnitude information can express richer texture information than that only using direction information. Therefore, the magnitude map is obtained from the magnitudes of horizontal and vertical as the fifth texture descriptor, which is expressed as:

$$MAG(p_c) = \sum_{i=1}^{8} 2^{i-1} \times \text{sgn}(M(p_i) - M(p_c)) \quad (6.16)$$

with

$$M(p_i) = \sqrt{\left[I^1_{0°}(p_i)\right]^2 + \left[I^1_{90°}(p_i)\right]^2} \qquad (6.17)$$

where p_i represents the neighboring pixel to p_c.

So far, multi-scale edge and texture features were extracted from the corresponding multi-scale monocular and cyclopean views, and then combined and mapped into a final global score assisted by a just noticeable difference (JND) model. The paper will be discussed again in detail about the mechanism of the JND in the next section.

6.4 SIQA Based on Visual Saliency

6.4.1 Visual Saliency Models

Visual attention, as one of the most important mechanisms of the HVS, has been mentioned in this chapter many times. In short, visual attention represents a process that enables the HVS to select the most relevant areas from a visual scene. The distinction between relevant and un-relevant areas can be mainly realized by two components: top-down factors driven by task and bottom-up factors driven by stimulus. More specifically, top-down attention deals with high-level cognitive factors (e.g., task demands, emotions, and expectations), expecting to distinguish the relevant regions. In contrast to top-down, bottom-up attention highlights image regions that are different from their surroundings by low-level visual stimulus, such as luminance, color, edge, orientation. In addition to the components of bottom-up and top-down, visual attention also includes other factors, such as overt/covert, spatial/spatio-temporal, and space-based/object-based attention. They all study and analyze the mechanism of visual attention through different entry points. Since our main purpose is not to introduce various visual attention concepts, we refer the interested reader to general reviews for full understanding about visual attention (Borji and Itti 2013; Borji et al. 2013).

Among the visual attention conveying various topics, due to the relative simplicity of bottom-up processing compared to other attention models, modeling the mechanism of bottom-up visual attention has been studied by numerous researchers and many successful models have been designed for the applications in image processing and computer vision. In general, the bottom-up visual attention, also known as visual saliency, has become a hot research direction in the field of image and computer vision.

The first well-known saliency detection model was perhaps proposed by (Itti et al. 1998), in which multiscale image features (e.g., image pixel intensity, orientation and color contrast) were combined into a single topographical saliency map. Then many improved saliency detection algorithms have been emerged in large numbers. (Koch and Ullman 1985) designed a visual saliency computational architecture by distinguishing the difference between the image region and its surroundings. In (Walther and Koch 2006), saliency toolbox was proposed as an extension for improving the output of Itti by extracting the region of interest. By modeling the behavior of HVS, (Le Meur et al. 2006) computed saliency using contrast sensitivity function, visual masking and the center-surround interactions simultaneously. According to the principle of maximizing the sampling information in a scene, (Bruce and Tsotsos 2009) designed an AIM model for saliency detection by using Shannon's self-information. Garcia-Diaz et al. (2012) designed an adaptive whitening saliency (AWS) model by considering the decorrelation of neural responses. In addition, based on the mathematical theories, (Hou and Zhang 2007) and (Guo and Zhang 2010) constructed saliency detection models respectively by exploring the phase spectrum of Fourier transform, namely SR and PQFT, which have proved to be simple yet efficient models. Seo and Milanfar (2009) constructed SDSR model by measuring the likeness of a specific pixel to its surroundings. Li et al. (2011) extracted global features from frequency domain and local features from spatial domain, and combined them into SDFS model.

Generally, the above-mentioned saliency detection models can be roughly divided into three categories according to the reference paper (Achanta et al. 2009). The first category of saliency models constructs the visual saliency model by extracting the underlying visual features based on the physical and psychological studies of the early primate visual system, such as Itti model. The second category of saliency models is driven by the mathematical computational methods and the practical need of object detection without any knowledge of the HVS, among which the classical examples are the saliency models including the SDSR and SR models. The last category combines the thoughts of the first two types of saliency models, considering both biological and mathematical models (e.g., the graph-based visual saliency (GBVS) model).

So far, numinous saliency models have been designed and applied in the fields likes object detection, among which GBVS and SR saliency models, as the most classical saliency detection models, will be introduced in the following in detail.

The models based on graph theory are a kind of saliency detection models, which have studied and developed in the past one decade. Among them, the GBVS model proposed in (Harel et al. 2007) is a typical model contributing to the major waves in the chronicle of saliency detection, which consists of three steps. The first step is to extract feature maps from the input image using the method similar to Itti model. For each extracted feature map M, corresponding activation maps can be formed from extracted feature maps according to the principle of Markov's method, which can be expressed by:

$$d((i, j)\|(p, q)) = \left|\log \frac{M(i, j)}{M(p, q)}\right| \qquad (6.18)$$

where $M(i, j)$ and $M(p, q)$ are the intensity values of nodes (i, j) and (p, q) in an extracted feature map, respectively. $d((i, j) \| (p, q))$ denotes the dissimilarity of $M(i, j)$ and $M(p, q)$.

6.4 SIQA Based on Visual Saliency

According to the graph theory, a fully-connected directed graph G_A is formed to connect every node in the input image with all the others. Then the direct edge can be assigned further a weight from node (i, j) to node (p, q) as following:

$$\omega_1((i, j), (p, q)) = d((i, j) \| (p, q)) \times \exp\left(-\frac{(i-p)^2 + (j-q)^2}{2\sigma^2}\right) \quad (6.19)$$

where σ is a free parameter.

So far, given a feature map, a directed fully connected graph can be obtained from it and its connection weights can also be clearly defined.

The last step is to normalize the activation maps A, which can be introduced in detail in following. Given A, every node is connected to all other nodes to construct graph G_N. Then the edge from node (i, j) to node (p, q) can be calculated by:

$$\omega_2((i, j), (p, q)) = A(p, q) \times \exp\left(-\frac{(i-p)^2 + (j-q)^2}{2\sigma^2}\right) \quad (6.20)$$

Now each edge weight is assigned a coefficient to make the sum of all edges equal to 1, which is called normalization step. Normalizing the weights of the outbound edges of each node boundary and treating the resulting graph as a Markov chain, the equilibrium distribution can then be calculated as a saliency map. Finally, multiple activation maps are fused to form a single saliency map.

As another different way to form saliency map, (Hou and Zhang 2007) proposed a saliency detection model by extracting spectral residual (SR) of the image in spectral domain and converting it into spatial domain, where the main construction steps can be formulated as:

$$\begin{aligned} A(f) &= \text{Re}\{F[I(x, y)]\} \\ P(f) &= An\{F[I(x, y)]\} \\ L(f) &= \log[A(f)] \\ R(f) &= L(f) - h_n(f) \otimes L(f) \end{aligned} \quad (6.21)$$

where F represents the Fourier transform. Re{} and An{} denote the operations of. A and P are the amplitude and phase spectrum of the image, respectively. R is the component of spectral redidual h_n denotes the local statistics neibor filter, which is described as follows:

$$h_n(f) = \frac{1}{n^2}\begin{pmatrix} 1 & 1 & \cdots & 1 \\ 1 & 1 & \cdots & 1 \\ \vdots & \vdots & \ddots & \vdots \\ 1 & 1 & \cdots & 1 \end{pmatrix} \quad (6.22)$$

Applying the inverse Fourier transform, the saliency model in spatial domain can be obtained by:

$$SR(x, y) = g(x, y) \otimes F^{-1}\{\exp[R(f) + iP(f)]\}^2 \quad (6.23)$$

where F^{-1} is the operation of inverse Fourier transform, and g denotes the Gaussian low pass filter for smoothing the saliency map for better visual effect:

$$g(x, y) = \frac{1}{\sqrt{2\pi\sigma^2}} \exp\left(-\frac{x^2 + y^2}{2\sigma^2}\right) \quad (6.24)$$

With the rapid development of stereoscopic image processing technologies, many researchers turn to the focus on stereo visual attention. Compared with 2D tasks, 3D saliency models much emphasize the visual perception of viewers, especially the binocular perceptual vision. So far, many computational visual attention models have been established for stereoscopic images and videos, which can roughly be classified into three types according to the computational thoughts. The first kind of the algorithms take into account the depth information by considering them as the weighting function of 2D saliency model. The models belonging to the second type incorporate the depth information into the traditional 2D saliency detection methods. The last way to establish 3D visual saliency model takes the stereo vision into consideration. In the subsequent section, we will introduce some classical saliency detection models for stereoscopic images and videos belonging to the three types of algorithms.

As the unique visual factor, the depth information is first taken into account for extending some 2D saliency detection models into 3D versions. There are two ways to realize the 3D visual saliency models: taking the depth information as weighting function and incorporating depth map and 2D saliency models. In the former, (Zhang et al. 2010) proposed a 3D visual saliency model for stereo videos by two steps: (1) multiple perceptual attributes were extracted from stereo videos including depth cues luminance, color, orientation and motion contrast. (2) a depth-based fusion model was used to integrate these features for the 3D saliency model generation, which was constructed by:

$$S_{SVA} = \Psi\left[D \times \left(k_s S_s + k_m S_m + k_D D - \sum_{uv \in \{sm, sD, mD\}} e_{uv} C_{uv}\right)\right] \quad (6.25)$$

where k_D, k_s and k_m are the weighting coefficients for depth map D, static saliency S_s and motion saliency S_m, respectively. C_{uv} denotes a correlated coefficient between u and v, which is defined as: $C_{sm} = \min(S_s, S_m)$, $C_{sD} = \min(S_s, D)$ and $C_{mD} = \min(S_m, D)$. e_{uv} is a weighted coefficient for the corresponding C_{uv}.

The 3D saliency detection model proposed in (Jiang et al. 2014) belonged to the latter, which considered simultaneously depth cues, center bias and 2D visual saliency model. By comparing the pixel of the disparity map with a given threshold, the image can be classified into the foreground and background regions. In addition,

6.4 SIQA Based on Visual Saliency

the effects of center bias were also considered in this algorithm. Central bias can be modeled by 2D Gaussian filter and it's center fixation of the filter is fixed on the center of the input image.

$$\text{CB}(x, y) = \exp\left\{-\left(\frac{(x-x_c)^2}{2\sigma_x^2} + \frac{(y-y_c)^2}{2\sigma_y^2}\right)\right\} \quad (6.26)$$

where x_c and y_c represent the center pixels of an image, respectively. σ_x^2 and σ_y^2 denote the variances along the two axes respectively.

The final 3D saliency map was described by the sum of a series of individual cues:

$$S_{3D} = \omega_1 S_{2D} + \omega_2 CB + \omega_3 FM + \omega_4 BM \quad (6.27)$$

where w_1 to w_4 are the weighting coefficients for each model, respectively. S_{2D} is the 2D saliency model calculated by the spectral residual algorithm (Hou and Zhang 2007). *FM* and *BM* represent the foreground and background regions of an image, respectively.

In addition to the two mentioned ways, there exists a saliency construction model considering the stereo vision characteristics instead of just taking the depth cues into consideration. When viewing a 3D image, the area of interest of human beings is slightly different from the 2D image because it is susceptible to the phenomenon such as binocular rivalry or binocular suppression. Therefore, it is necessary to quantify 3D visual saliency based on the visual mechanisms including binocular fusion and rivalry. For example, a visual saliency framework for stereo vision was built in (Bruce and Tsotsos 2005) by adding interpretive neuronal units into 2D saliency model. For fully considering the effects of binocular vision, (Nguyen et al. 2019) proposed a deep visual saliency model by extracting seven low-level features (contrast, luminance, and depth information) from stereopairs and concatenated them to adaptively learning stereo visual attention to human perception. In addition, we will also introduce some 3D saliency models in the following sections. For more state-of-the-art 3D visual saliency models, the readers interested of this can read some relevant materials.

6.4.2 Application of Visual Saliency in 3D IQA

Visual saliency, reflected in distorted images, demonstrates that the distortions occurred in salient regions are more annoying than that in non-salient regions, which derives the conclusion that visual saliency has great influences on the perceptual quality of nature images. Over the past decades, the effect of visual saliency in image quality has attracted more and more attention from IQA researchers. They are committed to applying visual saliency to all stages of the IQA algorithms for improving prediction performance. Significant progress has been made in IQA for plane natural images in the past few years. Inspired by this, it is necessary to explore the close relationship between visual saliency and image quality from 3D images.

Overall, visual saliency models can be applied into 3D IQA in three ways. The first and the most popular way is utilizing visual saliency map as weighting function in the stage of monocular views' feature fusion or quality score pooling. Motivated by the fact that the distortion of the image will affect the contents of corresponding saliency map, the second way uses the generated saliency map as feature maps, which can reflect the perceptual quality of distorted images. In addition, the last way starts the root of visual saliency and takes visual saliency as the enhancement and inhibition factor by multiplying the saliency map and corresponding map including the input image and extracted feature maps pixel by pixel to enhance and weaken corresponding locations of the image, which is called salience enhancement.

What we discussed in Chap. 4 is that the final objective score can be obtained by averaging the two monocular scores generated from the left and right views by directly applying 2D IQA models respectively. However, this method fails to achieve good performance in asymmetrically distorted stereoscopic images, which can be attributed to utilizing the average strategy that different distorted views have the same effect on the perceptual quality. Considering that visual saliency presents the degree of human eye attraction, the averaged saliency map can be utilized as a weighting factor to represent the importance of each view of stereo images in the stage of quality score pooling, which is expressed by:

$$Q = w_L \times Q_L + w_R \times Q_R \tag{6.28}$$

where Q_L and Q_R denote the local quality scores for the left and right views, respectively. ω_L and ω_R are the weighting factors to representing the importance of each view from a stereopair, which can be established from visual saliency map corresponding to the left and right images (S_L and S_R):

$$\omega_L = \frac{\sum_{i,j} S_L(i,j)}{\sum_{i,j} S_L(i,j) + \sum_{i,j} S_R(i,j) + \varepsilon}$$
$$\omega_R = \frac{\sum_{i,j} S_R(i,j)}{\sum_{i,j} S_L(i,j) + \sum_{i,j} S_R(i,j) + \varepsilon} \tag{6.29}$$

6.4 SIQA Based on Visual Saliency

where ε is a very small positive constant to avoid the denominator being equal to zero.

In another form of visual saliency as weighting function, visual saliency map can be also used to measure the relative importance of monocular image maps including monocular views of a stereopair and corresponding feature maps by utilizing a pixel-wise scheme. Supposing F_L and F_R are the monocular maps from a stereopair and S_L and S_R denote the corresponding saliency map of the left and right views, the weighting procedure for generating the intermediate map F_C can be illustrated in general in below:

$$F_C(i,j) = \frac{S_L(i,j) \times F_L(i,j) + S_R(i,j) \times F_R(i,j)}{\sum_{i,j} S_L(i,j) + \sum_{i,j} S_R(i,j) + \varepsilon} \quad (6.30)$$

where F_C is the quality map after the integration. We can see that Eq. (6.30) is very similar to the function for binocular combination that we have introduced in the previous chapter. The main difference between them is the selection of weighting responses, where the binocular combination model (Chen et al. 2013) selects energy responses as weighting function, while in Eq. (6.30), saliency map is chosen for integrating. Actually, there have been some attempts that the saliency map is utilized as the weight function of binocular combination for the two monocular views or feature maps, achieving quite good performance. The framework of a SIQA metric in this kind of saliency-assisted combination strategy is shown in Fig. 6.3.

For example, (Xu et al. 2017) proposed a NR SIQA metric by utilizing multi-scale saliency-guided feature consolidation. The multi-scale global and local features were extracted from the left and right views respectively, including spatial entropy and spectral entropy. And then the saliency model was utilized to fuse these features from monocular views into cyclopean features for simulating the binocular rivalry, which can be expressed by:

$$F_n = \frac{S_{L,n}}{S_{L,n} + S_{R,n}} \times F_{L,n} + \frac{S_{L,n}}{S_{L,n} + S_{R,n}} F_{R,n} \quad (6.31)$$

where $S_{L,n}$ and $S_{R,n}$ denote multi-scale visual saliency of each view from a stereopair, respectively. Similarly, $F_{L,n}$ and $F_{R,n}$ are multi-scale global and local statistical features for each view. F_n represents the visual features after the procedure of feature fusion.

The second way applying visual saliency into IQA is to take saliency map as quality-aware feature maps. The relationship between visual saliency and image quality has been investigated by some researchers in previous studies and the fact that the regions that attract more attention will be change when distortion occurs in pristine images has been widely accepted in IQA fields. Thus, adding visual saliency for representing the perceptual quality of distorted images appropriately can benefit IQA metrics and could achieved better prediction performance.

There has been many research considering visual saliency as feature maps for characterizing the stereo image's local quality in 3D IQA. In (Yao et al. 2017),

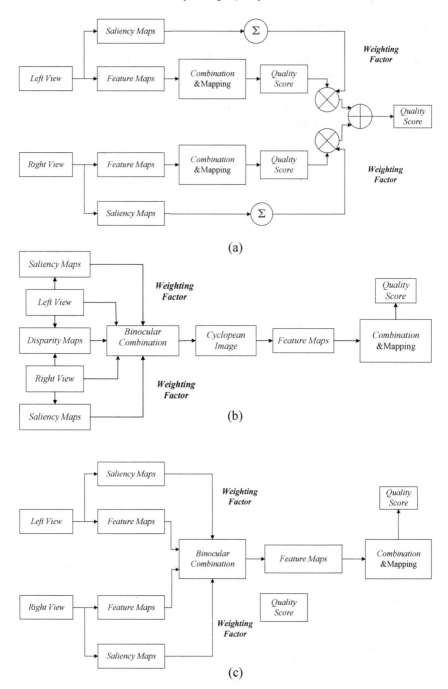

Fig. 6.3 The framework of a SIQA method using saliency map as weighting function for binocular combination: **a** quality score integration, **b** cyclopean image generation and **c** the generation of intermediate feature maps

6.4 SIQA Based on Visual Saliency

the authors proposed a FR SIQA metric, in which four kinds of visual features were extracted, including the similarity of visual saliency (VS) maps and gradient magnitude (GM) between the reference and distorted 3D images, the difference of binocular energy and the features from disparity map. The saliency detection model employed in this metric is SDSP proposed by Zhang et al. (2013), which has been introduced in the previous section. The similarity between the reference and distorted VS maps (S_r and S_d) is defined as:

$$\text{SM}_S(i,j) = \frac{2S_r(i,j)S_d(i,j) + C_1}{S_r(i,j)^2 + S_d(i,j)^2 + C_1} \quad (6.32)$$

where SM_S is the similarity map between S_r and S_d. C_1 denotes a small positive constant to avoid the denominator being zero for stability improvement.

Another insight is to extract more specific and robust visual features on saliency maps. (Yang et al. 2018) proposed a quality assessment algorithm for stereoscopic video by considering various information from saliency and sparsity. First, the authors established the 3D saliency map from the sum maps of video's frames using the saliency detection model proposed in (Fang et al. 2014), which was considered a valid visual map remaining the basic information of stereoscopic video. Second, sparse representation was utilized to decompose the 3D saliency map into sparse coefficients and derived the features from the 3D saliency map. The extracted features using the sparse representation method can reflect the visual characteristics of visual attention on the video's frames effectively. Finally, the extracted features were transformed into a series of sparse features through a stacked auto-encoder, and mapped into a final score using support vector regression. The general framework of the strategy taking visual saliency as quality-aware feature maps can be illustrated as following (Fig. 6.4):

The application of visual saliency in 3D IQA is not limited within the aforementioned ways. The idea of saliency enhancement has been adopted in many SIQA research and achieving pretty good performance, in which the main difference is the generation method of saliency map and the extracted feature maps (Fig. 6.5).

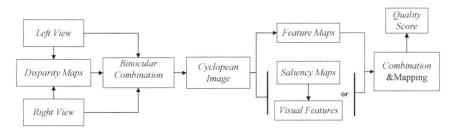

Fig. 6.4 The general framework of 3D IQA method where visual saliency map directly serves as feature maps or effective visual features are extracted from the saliency map

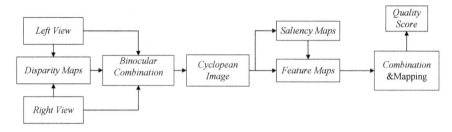

Fig. 6.5 The general framework of 3D IQA method where visual saliency map serves as a weighting function for enhancing and weakening the locations of stereo image

For example, (Zhang et al. 2016) first designed a 3D saliency detection model S_{3D} by integrating 2D GBVS map S_{2D} and the relative disparity map d.

$$S_{3D} = (1 - \alpha) \times S_{2D} + \alpha \times d \tag{6.33}$$

where the parameter α is a weighting factor that controls the relative importance between 2D saliency map and disparity map. Then, a series of disparity statistics including magnitude, contrast, dispersion, skewness and the average of maximum and minimum $p\%$ disparity values are given by Eqs. (6.34)–(6.37) from the saliency-enhanced disparity map for pooling the final objective score.

(a) 3D saliency-enhanced disparity magnitude:

$$f_1 = \frac{1}{d_m} \times \frac{\sum_{i,j} S_{3D}(i, j) \times |d(i, j)|}{\sum_{i,j} S_{3D}(i, j)} \tag{6.34}$$

where $|d(i,j)|$ is the magnitude map of disparity map. d_m represents the maximum disparity magnitude as a normalized factor.

(b) 3D saliency-enhanced disparity contrast:

$$f_2 = \frac{1}{d_m} \times \frac{\sum_{i,j} S_{3D}(i, j) \times |d_c(i, j)|}{\sum_{i,j} S_{3D}(i, j)} \tag{6.35}$$

where $|d_c(i, j)|$ is the disparity contrast map computed by using a center-surrounding operator.

(c) 3D saliency-enhanced disparity dispersion:

$$f_3 = \frac{1}{d_m} \times \sqrt{\frac{\sum_{i,j} S_{3D}(i, j) \times d(i, j)^2}{\sum_{i,j} S_{3D}(i, j)}} \tag{6.36}$$

6.4 SIQA Based on Visual Saliency

(d) 3D saliency-enhanced disparity skewness:

$$f_4 = \frac{\sum_{i,j} S_{3D}(i,j) \times |d(i,j)|^3}{\left[\sum_{i,j} S_{3D}(i,j) \times |d(i,j)|^2\right]^{3/2}} \qquad (6.37)$$

In another example adopting the saliency-assisted enhancement strategy (Wang et al. 2018), the authors designed a novel visual saliency model for stereoscopic images by considering the disparity map and difference image obtained from the stereopairs. Firstly, a new quaternion representation (QR) of each view from stereopairs was constructed from both the image content and disparity perspectives. The generation of QR can be expressed as Eq. (6.4).

$$I_q = I_i + \mu_1 I_C + \mu_2 I_d + \mu_3 M_d \qquad (6.38)$$

where u_i represents the unit coordinate vectors that are perpendicular to each other. I_i and I_C denote the image luminance and chrominance components from the left or right image of stereopairs, respectively. M_d and I_d are the disparity map and the difference map between the left and right views of stereopairs. I_q is the QR of each view.

Employing a phase spectrum likely (SR) saliency generation method already introduced in the previous section, visual saliency map can be obtained from the synthesized QR image. On the basis of generating stereoscopic visual saliency, the authors applied them into SIQA tasks for guiding the quality pooling stage to improve the performance. More specifically, the error maps from the left and right images can be generated by using 2D IQA metric, while the visual saliency maps of each view for stereopairs were obtained from corresponding QR images. By utilizing the pooling function listed in Eq. (6.39), the saliency-enhanced feature map of each view can be obtained.

$$F'(i,j) = \frac{S(i,j) \times F(i,j)}{\sum_{i,j} S(i,j)} \qquad (6.39)$$

where F represents the error map here. Finally, the overall score was given by:

$$Q = w_L \sum_{i,j} F'_L(i,j) + w_R \sum_{i,j} F'_R(i,j) \qquad (6.40)$$

where ω_L and ω_R are the weighting factors of the left and right views, respectively.

The third FR SIQA example adopting the weighting strategy is proposed by Yang (2016). The authors proposed a cyclopean and saliency-based IQA method for stereoscopic images, where the cyclopean saliency map and cyclopean image of stereopairs were calculated in a similar way as follows:

$$C(i, j) = f_B(I_L(i, j), I_R(i, j+d))$$
$$CS(i, j) = f_B(S_L(i, j), S_R(i, j+d)) \quad (6.41)$$

where C and CS represent the synthesized cyclopean image and cyclopean saliency map, respectively. f_B denotes a binocular combination model for synthesizing the cyclopean image and cyclopean saliency. To discuss and analyze the effects of binocular combination models, four common combination metrics were introduced, including eye-weighting model, vector summation model, neural network model and gain-control model, which have been introduced and discuss in Chap. 5.

After generating the cyclopean image and cyclopean saliency map from reference and distorted stereopairs, the authors considered the cyclopean saliency map as a weighting function to highlight the salient areas, which was realized by multiplying the cyclopean saliency map and the corresponding cyclopean saliency map:

$$CW_r(i, j) = C_r(i, j) \times CS(i, j)$$
$$CW_d(i, j) = C_d(i, j) \times CS(i, j) \quad (6.42)$$

where CW_r and CW_d are the reference and distorted cyclopean images after saliency-assist enhancement, respectively. Last, the obdsjective quality of the distorted stereopairs was obtained by using a FR IQA algorithm for 2D images to CW_r and CW_d, which can be formulated as:

$$Q = f(CW_r, CW_d) \quad (6.43)$$

where Q is the overall quality and f represents the FR IQA metric for 2D images. The overall framework is illustrated as Fig. 6.6.

The last example we would like to introduce is a FR 3D IQA metric (Liu 2016). The authors started from the root of input images and multiplied the generated 3D saliency map and corresponding cyclopean image pixel by pixel. The 3D saliency map S_f was first established according to the saliency detection idea proposed in (Fang et al. 2013). Considering the effect of the center bias factor and normalized visual sensitivity $C_s(f, e)$ (Geisler and Perry 1998a, b), the final 3D saliency map can be expressed as:

$$S_{3D} = (0.7 S_f + 0.3 S_c) \times C_s(f, e) \quad (6.44)$$

where S_c represents the center-bias map calculated by the center bias factor.

For a stereopair including the left and right views, the synthesized cyclopean image can be obtained by adopting four common binocular combination models discussed in last Chapter. Let the cyclopean image and its corresponding saliency map of the reference stereopair named C_r and $S_{3D,r}$, and meanwhile the distorted versions denoted as C_d and $S_{3D,d}$. The cyclopean image after saliency-guided enhancement (C'_r and C'_d) can be built as:

6.4 SIQA Based on Visual Saliency

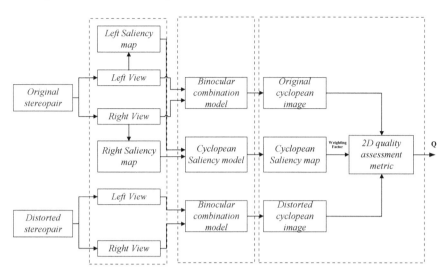

Fig. 6.6 The framework of FR SIQA method proposed in Yang (2016)

$$C'_r(i, j) = C_r(i, j) \times (1 + \alpha \times S_{3D,r}(i, j))$$
$$C'_d(i, j) = C_d(i, j) \times (1 + \alpha \times S_{3D,d}(i, j)) \quad (6.45)$$

where α is a positive to highlight the importance of visual saliency on image enhancement. Finally, the task of the FR 3D IQA can be transformed into 2D IQA task, which can be solved by applying some well-known IQA metrics for plane images.

In fact, researchers sometimes may apply visual saliency into IQA tasks not just with a single saliency-assisted strategy, while with various saliency strategies together discussed above to improve performance. Here we would like to conclude this section by introducing a classical 2D FR IQA metric that takes advantage of visual saliency (Zhang et al. 2014) to end this section, although it does not belong to the field of 3D IQA. In this paper, visual saliency was considered a feature map characterizing the quality of images, as well a weighting function in the stage of quality score pooling. First, both visual saliency map and gradient magnitude were extracted as two basic feature maps. After calculating their structural similarity separately, the two similarity maps of saliency and gradient magnitude were combined to obtain the objective quality, where the similarity of between saliency maps and that between gradient maps are formulated as following:

$$\text{SIM}_V(i, j) = \frac{2V_r(i, j) \times V_d(i, j) + C}{V_r(i, j)^2 + V_d(i, j)^2 + C} \quad (6.46)$$

where V_r and V_d denote the saliency or gradient map corresponding to the reference and distorted images. C is a positive constant for avoiding the denominator being equal to zero. In addition, visual saliency also serves as a weight matrix to derive the final score by the following function:

$$\text{VSI} = \frac{\sum_{i,j} SIM(i,j) \times W(i,j)}{\sum_{i,j} W(i,j)} \tag{6.47}$$

where W represents the weighting factors obtained from the maximum between the reference and distorted saliency maps.

6.5 SIQA Based on Just Noticeable Difference

6.5.1 Just Noticeable Difference

Research on psychophy and physiology has found the phenomenon that the human visual system (HVS) is only capable of perceiving the image pixel change above a certain visibility threshold, while HVS is insensitive to the image areas below the visibility threshold. The above description is another important visual property of HVS, also called just noticeable difference (JND). It was first proposed in Weber-Fechner law (Boring 1944), and improved by the subsequent research (Barlow 1957), which can be defined by:

$$\frac{\Delta I}{I} = \text{const.} \tag{6.48}$$

where I is the original luminance, and ΔI represents the required noticeable change over the luminance. const. is a constant called Weber fraction.

According to the definition of JND, how to determine the threshold is a puzzled work, that's because relatively slighter distortion will be missed when setting a high threshold, while a low threshold will reduce the effect of JND on content's detection. We have known that whether the changes of image can be perceived by human beings is determined by the property of JND, which can be recognized as an important visual factor for watermarking (Bouchakour et al. 2008), image enhancement (Huang et al. 2008) and image quality evaluation (Toprak and Yalman 2017), especially for the quality evaluation of distorted images. Taking this property into account is very useful for improving the performance of IQA prediction, as it allows small undetected distortions to be ignored, which is highly consistent with HVS. Thus, correctly understanding and applying the characteristic of JND is crucial when we want to best simulate the physiological and psychological mechanisms of human perception of images. On the other hand, because the HVS is the final receiver of image information, understanding and studying the visual mechanisms of HVS is of vital importance for deploying a more reliable JND models. Here we will introduce some visual masking effects that are critical for the JND model, including spatial contrast sensitivity, luminance adaptation, contrast masking and temporal masking.

Masking effect is a complicated perceptual phenomenon that the visibility reduction of one signal to human eyes in the presence of another signal when these signals

6.5 SIQA Based on Just Noticeable Difference

occur in space simultaneously. For 2D images, masking effect is mainly affected by spatial masking and temporal masking, in which the spatial masking effect can be further classified into two effects: luminance masking and contrast masking. According to Weber-Fechner law, the perceptible luminance difference of a stimulus depends on the surrounding luminance level, which means human eyes are more sensitive to luminance contrast rather than the absolute luminance value. A simple single-stimulus image I_t can be expressed as the surrounding background luminance I_b plus the luminance difference I_d:

$$I_t = I_b + I_d \tag{6.49}$$

When the scene with the background luminance I_b of high levels, the luminance contrast remains nearly a constant with the background luminance slightly increasing or decreasing. On the contrary, the lower background luminance is likely to result in the luminance contrast varying with the background luminance. This can be described by the fact that a high visibility threshold of luminance contrast needs to be set in dark regions, while a lower visibility threshold in bright areas.

Psychophysical research has found the phenomenon that the HVS can tolerate more noises in textured regions than other smooth regions since frequent spatial activities decrease the capability of the image difference detection, which is caused by the contrast masking. The contrast masking can be defined as the spatial inhomogeneity of the background brightness resulting in the reduction in the ability of the visibility of the stimulus.

There exists a special masking effect for videos, named the temporal masking. This is based in the principle that visual contents vary over time. Since the target of our research is mainly digital image, the detailed discussion about the temporal masking will not be given in this book.

The research on JND first made a breakthrough in plane image. According to the domain for the visibility threshold being calculated, JND models can be classified into two categories roughly: spatial domain and frequency domain. Typical spatial domain JND models taking spatial masking effects into account, while the frequency domain JND models consider temporal masking effects and measure the JND threshold in frequency domain such as sub-band (Safranek and Johnston 1989; Jia et al. 2006), discrete cosine transform (DCT) (Peterson et al. 1991; Hahn and Mathews 1998) and wavelet domains (Watson et al. 1997; Wei and Ngan 2009). In spatial domain, several JND models have been proposed to simulate luminance masking, contrast masking or their combination. Chou and Li (1995) first proposed a JND model by catering to the dominant relationship between the background luminance and the luminance contrast (or namely Weber fraction) and then determining the visibility threshold of luminance masking effects according to the average luminance of nearby pixels of a certain point, which is a simple but effective model for quantifying perceptual redundancies. According to the experimental results, the visibility threshold corresponding to low background luminance regions (i.e., less than 127) was modelled by a root equation, while in the other regions, the visibility threshold was approximated by a linear function, as equivalently described as follows:

$$T_l(i, j) = \begin{cases} 17\left(1 - \sqrt{\frac{L(i,j)}{127}}\right) + 3, & \text{if } L(i, j) \leq 127 \\ \frac{3}{128}(L(i, j) - 127) + 3, & \text{otherwise} \end{cases} \quad (6.50)$$

where $T_l(i,j)$ denotes the visibility threshold of luminance masking effects. $L(i,j)$ is the background luminance of the image $I(i,j)$, which can be calculated by:

$$L(i, j) = \frac{1}{32} \sum_{x=1}^{5} \sum_{y=1}^{5} I(i - 3 + x, j - 3 + y) \times F(x, y) \quad (6.51)$$

where $F(i,j)$ represents a weighted low-pass filter with the size of 5×5.

In addition, the effect of contrast masking can be modelled by a function of the background luminance and the contrast of a certain pixel, which was expressed as follows:

$$T_c(i, j) = c_1 L(i, j) \times (c_2 G(i, j) - 1) + c_3 G(i, j) + c_4 \quad (6.52)$$

where c_1 to c_4 are constants for adjusting the weights of each part. G represents the weighted average of gradient around the pixel (i, j).

This approach overestimates the visibility threshold of contrast masking for edge regions, which can be attributed to the distortion in edge regions will attracts more visual attention than that in textured regions. Thus, (Yang et al. 2005) divided the effect of contrast masking into edge masking and texture masking, and proposed a nonlinear additivity model for masking (NAMM) by counting for the effects of the luminance masking and texture masking with a provision to deduct their overlapping effect, which can be expressed by:

$$T(i, j) = T_l(i, j) + T_t(i, j) - C(i, j) \times \min\{T_l(i, j), T_t(i, j)\} \quad (6.53)$$

where T_l and T_t are the visibility thresholds for the two primary masking factors: luminance masking and texture masking. C accounts for the overlapping effect of masking ($0 < C \leq 1$). The luminance masking model proposed in Eq. (6.50) is also applied here, and meanwhile the texture masking can be determined with local spatial activities such as gradients around the pixel, which is defined as follows:

$$T_t(i, j) = \eta \times G(i, j) \times W_e(i, j) \quad (6.54)$$

where η is a control parameter, and W_e denotes an edge-related weight of the pixel (i, j) calculated by Canny's detector.

Similarly, (Liu et al. 2010) improved the NAMM by considering the effects of both the edge masking and texture masking. The targeted image was first decomposed into structural and textural regions, respectively. The visibility threshold of contrast masking can be obtained from the structural and textural regions:

6.5 SIQA Based on Just Noticeable Difference

$$T_c(i, j) = \eta \times (w_e T_e(i, j) + w_t T_t(i, j)) \tag{6.55}$$

where T_e and T_t represent the visibility thresholds of edge masking and texture masking that calculated by Eq. (6.54) for both structural and textural images, respectively. w_e and w_t are the weights for edge masking and texture masking, representing the significance of each masking effect, respectively.

If the change of luminance difference is, for example, a sine wave, the visibility threshold will also be influenced by spatial and temporal frequencies of I_d, which is known as the contrast sensitivity function (CSF). Unfortunately, the pixel-wise JND models in spatial domain don't take into account the CSF that is one of the major factors affecting the JND mechanism. Thus, this kind of pixel-wise JND models cannot simulate the HVS completely. There is another kind of JND models operating in frequency domain that can easily incorporate the CSF into the JND profile. The detailed implementation is that an image is first transformed into frequency domain before all the subsequent operations. In (Safranek and Johnston 1989), the JND map was produced from 16 separated sub-bands that were generated by some spatial filter banks and pyramid decomposition. In addition, JND also can be measured in DCT domain, which is realized by dividing an image into blocks of size $N \times N$, and transforming them into DCT domain. The DCT-based JND can be described as:

$$JND_{DCT}(i, m, n) = T_{CSF}(i, m, n) \prod_j \eta_j(i, m, n) \tag{6.56}$$

where T_{CSF} is the base threshold that is related to the spatial CSF. η_j denotes the parameter result from the j-th factor. JND_{DCT} represents the JND map in DCT domain. I denotes the ordinal of the image blocks, and (m, n) is a DCT sub-band ($m, n = 0, 1, 2, \ldots, N - 1$).

With the development of stereo image and video technologies, stereo vision become a hotspot in the field of image processing. However, the above mentioned JND models are generally based on specific characteristics of monocular vision, which are not applicable to the complex stereo perception of both monocular and binocular visual cues. In addition to the 2D masking effects that have been introduced before, the depth masking effect need to be considered for 3D images, which has been studied and explored in (De Silva et al. 2010, 2011). For 3D images/videos, the small changes of the depth perception on the scene are hardly perceived by the HVS. In the other words, the small changes of the depth perception below a visibility threshold can hardly affect the quality of the 3D images.

To better fit with the complexity of 3D visual perception, the concept of JND has been modified to adapt to stereo vision in recent years, called 3D JND. still, only few 3D JND models have been proposed because the binocular contrast masking behavior is a very complex visual process compared to its 2D counterpart, which brings to a large challenge for describing and simulating the property of 3D JND. A widespread 3D JND model (Zhao et al. 2011), called binocular JND (BJND), revealed the relationship between the visibility threshold and binocular vision according to a series of psychophysical experiments. Based on these experiments, the mathematical BJND model is designed by jointly considering the binocular combination of noises and the reduction of visual sensitivity in binocular vision. Note that, there is a corresponding BJND threshold for the left and right views of a stereopair, respectively. The generation of the two BJND threshold adopts the same approach, which is defined by:

$$BJND_l = BJND_l(bg_r(i+d,j), eh_r(i+d,j), A_r(i+d,j))$$
$$= A_{C,\text{limit}}(bg_r(i+d,j), eh_l(i+d,j))$$
$$\times \left[1 - \left[\frac{A_r(i+d,j)}{A_{C,\text{limit}}(bg_r(i+d,j), eh_l(i+d,j))}\right]^\lambda\right]^{1/\lambda} \quad (6.57)$$

where d is the disparity value at pixel (i, j). bg_r represents the background luminance of the right view, and eh_v denotes the edge height of the left or right view, where $v \in \{l, r\}$. A_r is the noise amplitude and λ denotes a constant. $A_{C,\text{limit}}$ denotes the visibility thresholds affected by the contrast masking.

To consider the effects of the depth masking, a just noticeable difference in depth (JNDD) was proposed and developed in (De Silva et al. 2010, 2011). (De Silva et al. 2010) recognized that the visibility threshold of the JNDD was mainly influenced by the viewing distance and the displayed depth level of the images, and established a JNDD model as follows:

$$JNDD = 10^{[0.94 \times \log_{10}(v) - 2.25]} + K_w \times |dp| \quad (6.58)$$

where v represents the viewing distance between the viewer's eyes and the screen. dp denotes the simulated depth level with meter unit, and K_w is the Weber constant.

Another 3D JND model considering the depth perception was proposed by Li et al. (2011), namely joint JND (JJND) model. The JJND model took into account the disparity estimation, image decomposition and JND threshold distribution. Firstly, disparity estimation was performed to distinguish between the occlusion region (OR) and non-overlapped regions (NOR) of a 3D image. Then, different JND thresholds were applied on different regions through the different generation strategy as follows:

$$JJND(i,j) = \begin{cases} T(i,j) \times \alpha, & \text{if } I_r(i,j) \in OR \\ T(i,j) \times \beta(i,j), & \text{otherwise} \end{cases} \quad (6.59)$$

where $T(i, j)$ is the visibility threshold for monocular images using the NAMM model. α denotes the depth perception parameter, and β is the joint masking effect parameter, which is determined by the depth of the image.

For better representing the stereopairs with low depth perception, (Zhou 2012) designed a multi-view JND (MJND) model by combining the spatial masking, temporal masking and depth masking, as depicted in below:

6.5 SIQA Based on Just Noticeable Difference

$$MJND(i, j) = T_s(i, j)^{w_1} \times T_{te}(i, j)^{w_1} \times T_d(i, j)^{w_1} \tag{6.60}$$

where T_s, T_{te} and T_d represents the visibility threshold affected by the spatial masking, temporal masking and depth masking, respectively. w_1 to w_3 are the weighting parameters for the three masking effects.

In addition to the previously mentioned 3D JND models, there are also some relatively new 3D JND models that are proposed and released in recent research. For example, (Qi et al. 2013a, b) proposed a stereo JND model (SJND) for stereo videos by considering the masking effects of both the intra-view and inter-view. For better representing the depth perception, (Zhong et al. 2012) developed a new 3D JND model by combining 2D JND models with the depth saliency that was generated using the depth intensity and depth contrast. This model was further proved in (Zhong et al. 2015), namely the hybrid JND (HJND) model, by additional consideration of the effects of geometric distortion. In (Du et al. 2016), to uncover the relationship between the visibility threshold and texture complexity, an asymmetrically distorted stereoscopic image database with different texture densities was generated for the subjective experiments, and a non-linear 3D JND model was designed for modelling the effect of texture complexity on the visibility threshold.

6.5.2 Application of JND in 3D IQA

The property of the JND has shown remarkable ability in terms of whether or not the changes of image content are perceived by human eyes, which is recognized as an important visual factor in the HVS. Since the HVS is the receiver of the image information, taking into account the JND in IQA tasks could improving the prediction performance.

A general insight to apply the property of the JND into IQA tasks is based on the fact that a pixel with high JND threshold means this pixel can tolerate large pixel changes, and further demonstrates the low importance of this pixel in the corresponding image. Therefore, like visual saliency, the JND can be applied as a weighting function to enhance or weaken the pixel of a stereopair:

$$Q = \frac{\sum_{i,j} \frac{1}{JND(i,j)} \times F(i, j)}{\sum_{i,j} \frac{1}{JND(i,j)}} \tag{6.61}$$

where F represents the extracted feature map, and Q means the integrated quality score.

Based on this thought, (Shao et al. 2013) utilized the BJND model (Zhao et al. 2011) to reflect the visual sensitivity, and to calculate the integrated quality scores for the binocular fusion and suppression regions, respectively. In (Fezza et al. 2014), the authors divided the stereopair into occluded and non-occluded regions, in which a JND model (Liu et al. 2010) was utilized to adjust the quality score of the occluded regions, and meanwhile the BJND model (Zhao et al. 2011) was used to modulate the quality scores of the non-occluded ones. The extracted feature map was generated by the structural similarity (SSIM) from the left and right views. In addition to the strategy of JND-weighting, (Fan et al. 2017) combined the quality of the JND-based cyclopean map with the quality of disparity map to assess the final quality score. Gu et al. (2019) first generated multi-scale intermediate left and right images using log-Gabor wavelet and the corresponding disparity maps generated by a SSIM-based stereo matching algorithm. Then multi-scale cyclopean images were established by the binocular fusion model that has been introduced in Chap. 5. Then edge and texture features were extracted from those synthesized monocular and binocular images. Different multi-scale 2D and 3D JND maps were considered in the integration procedure for different image targets. Specifically, a 2D JND model (Liu et al. 2010) was adopted as a weighting function to adjust the quality score for monocular images including the left and right views, the BJND model for cyclopean images and the JNDD model (De Silva et al. 2010) for the disparity maps. Finally, these adjusted multi-scale monocular and binocular features were pooled into a global score using support vector regression (SVR).

The JND can reflect the importance of the pixel from the stereopair by observing the maximum tolerance of the changes in one pixel of a stereopair. Similar to visual saliency, the JND map also can be recognized as a special feature maps to reflect the visual attention of the HVS. For example, (Qi et al. 2013a, b) developed a full-reference quality assessment algorithm for stereo videos by capturing spatio-temporal distortions and binocular perceptions, in which a SJND model was proposed for generating relative feature maps. Then the similarity maps between the reference and distorted feature maps can be obtained by:

$$q(i, j, t) = \frac{2 SJND_r(i, j, t) \times SJND_d(i, j, t) + \varepsilon}{SJND_r^2(i, j, t) + SJND_d^2(i, j, t) + \varepsilon} \quad (6.62)$$

where $SJND_r$ and $SJND_t$ are the SJND maps generated from the reference and distorted t-th frame of a stereo video, respectively. ε denotes a small positive constant for avoiding the denominator being zero. The final global score was generated by the sum of the similarity maps along to the times.

Psychophysical experiments have demonstrated that visual saliency highlights the regions attracted by the HVS, while the JND represents the visual perception threshold within the salient regions. Taking both visual saliency and JND into account could improve the prediction performance of 2D and 3D IQA tasks. By the approaches of permutation and combination, there are four general strategy for combining visual saliency and JND model, as depicted in Fig. 6.7.

For example, the model proposed in (Qi et al. 2015) was an improved version of (Qi et al. 2013a, b) by additionally adding the property of visual attention. The authors established a binocular visual saliency model as a weighting function to modulate the quality score generated by the SJND maps, which can be expressed by:

6.5 SIQA Based on Just Noticeable Difference

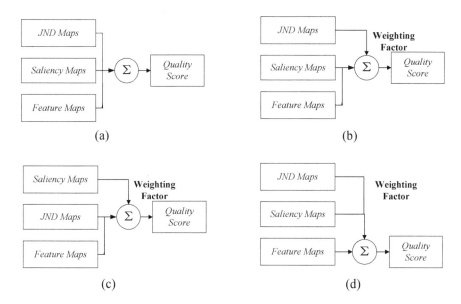

Fig. 6.7 Four general framework for combining visual saliency and JND model in IQA tasks. **a** Taking both visual saliency and JND map as feature maps, **b** JND model for weighting function and saliency map for feature maps, **c** saliency map for weighting function and JND model for feature maps, and **d** combining the saliency map and JND model as weighting function

$$Q = \sum_{i,j,t} q(i,j,t) \times [\omega_{3D} \times S_F(i,j,t) + (1 - \omega_{3D}) \times (1 - S_F(i,j,t))] \quad (6.63)$$

where S_F represents the designed binocular saliency model, and w_{3D} is the weight of salient region. Another example we would like to introduce was (Li 2019), in which the authors combined visual saliency and JND to simulate visual attention and perception for weighting the visual features. The weighting strategy can be described as follows:

$$F_C(i,j) = \omega_1(i,j) \times \omega_2(i,j) \times F_L(i,j) + (1 - \omega_1(i,j) \times \omega_2(i,j)) \times F_R(i,j) \quad (6.64)$$

with

$$\omega_1(i,j) = \frac{S_L(i,j)}{S_L(i,j) + S_R(i,j)}, \quad \omega_2(i,j) = 1 - \frac{J_L(i,j)}{J_L(i,j) + J_R(i,j)} \quad (6.65)$$

where S_L and S_R are the saliency map calculated from the left and right views by using a spectral residual approach (Hou and Zhang 2007). J_L and J_R denote the JND models for monocular views proposed in (Liu et al. 2010). F_L and F_R represents the local and global feature maps extracted from the left and right images, and F_C is the synthesized cyclopean feature map.

In addition to the aforementioned applications, the property of the JND determines it also benefits the image partition. Since local high contrasted regions in one view tends to suppress the regions with low contrast in the other view, namely binocular rivalry, the inter-difference visibility threshold and the local contrast comparison between the two views can model the mechanism of binocular vision. By monitoring whether or not binocular rivalry may occur in specific regions according to the comparison between pixel distortion and its corresponding JND threshold, the stereopair could be segmented into four disjoint regions for each view, which was proposed in (Hachicha et al. 2013). The authors defined four different regions including occlusion, invisible distortion, binocular suppression and binocular rivalry regions, as follows:

(1) Occlusion region contains occluded pixels in the monocular view plus the associated pool of pixels offset by zero parallax and overflowed differences:

$$O_v = \{(i, j) \in v, d_v(i, j) = 0\} \cup \{(i, j) \in v, i - d_v(i, j) < 0\} \quad (6.66)$$

where d_v is the disparity map corresponding to the left or right view, where $v \in \{l, r\}$.

(2) Invisible distortion region consists of the non-occluded regions with low contrast, where the change of the pixel cannot be recognized by human observers:

$$T_v = \{(i, j) \in \overline{O_v}, |\Delta I_v(i, j)| < BJND_v(i, j)\} \quad (6.67)$$

where ΔI represents the contrast changes of monocular view, and *BJND* is the adopted BJND model proposed in (Zhao et al. 2011). $\sim O_v$ denotes the non-occlusion region of monocular view.

(3) Binocular suppression region represents the pixel that are non-occluded and meanwhile satisfy the principle that the inter-difference between the two views is not visible, i.e., it is less than the BJND threshold.

$$S_v = \overline{O_v} \cap \overline{T_v} \cap C_v \cap D_v \quad (6.68)$$

where C_v means the left to right local contrast comparison criteria, and D_v is the principle that the inter-difference between the two views is less than the BJND threshold. For example, C_l and D_l can be expressed by:

$$C_l = \{(i, j) \in l, \text{LC}_l(i, j) > \text{LC}_r(i, j)\} \quad (6.69)$$

$$D_l = \{(i, j) \in \overline{O_l}, |I_l(i, j) - I_r(i, j)| < BJND_l(i, j)\} \quad (6.70)$$

where LC_l represents the local contrast of the left view.

(4) In contrast to binocular suppression region, binocular rivalry region is the region that the inter-view difference is visible and exceeds the BJND threshold.

$$R_v = \overline{O_v} \cap \overline{T_v} \cap C_v \cap \overline{D_v} \tag{6.71}$$

There are different methods to calculate the local quality score for different regions. Finally, the local quality scores of the four regions were aggregated into the final quality score.

6.6 Summary

This chapter has been discussing the properties of HVS that are or potentially are useful for designing IQA methods, and selected examples of successful IQA methods making use of HVS characteristics are introduced to highlight the effectiveness of this designing approach. In the history of the last few decades, discoveries in visual neuroscience and the applications in engineering including image processing and computer vision have influenced, interacted and inspired each other. The findings in neuroscience has given lots of novel ideas of IQA study as well as proofs for the theoretical ground of IQA methods. Although it has to be admitted that HVS is still yet to be thoroughly studied, we can safely state that studying IQA from the prospective of HVS sensation has become completely feasible.

Firstly, the basic structure of HVS is known to us. In the past, researchers used to take HVS as a black-box and study its properties regardless of its inner structure. This is certainly due to the limitation of neurobiological findings. Nowadays, we have been aware that the topological structure that HVS has is basically hierarchical, rather than flat, as it was once assumed, or any other form. The hierarchies exist in both the visual cortex, which is the region in cortex in charge of visual signal processing, and the regions before that, which are composed by the eyes and the connections between the eyes and the visual cortex. According to this hierarchical structure, we know that our ability with vision, including the ability to judge image quality, to detect and recognize things and faces, to track the movement of objects, etc., is the accumulation of processing through multiple stages. Concretely, the cells in our eyes are categorized into two kinds, those to sense luminance and those to sense color. The cells are operating independently upon each other, yet the signals they carry are integrated and edges, bars, and spatial frequencies are perceived by simple cells in the first visual cortex. Later in the second visual cortex, shapes and textures are composed. In further stages, or in the visual cortex regions that are further from the visual signal receptors, the visual signals are gradually more and more abstracted. Finally, we can let the simple luminance and color signals compose objects and see if they match certain things in our memory (recognition) and detect and tract their locations (localization). HVS is a highly mature system that has been evolved to be very effective for visual signal processing, so, it is reasonable to make use of its hierarchical structure to accomplish the compute vision tasks that are easy for human. Actually, the hierarchical structure is employed for human neural system for other purposes other than generating vision. The success and popularization of neural network is a good example of modeling this hierarchy. Focusing on IQA, the modeling can be simplified because low-level features that are attractive for the eyes and early visual cortex regions are enough for quality description, unlike highly abstracted goals such as detection, recognition, etc. Therefore, simulating HVS by constructing cascaded procedures to model its different stages can be feasible for IQA. Problem is, the present understanding upon HVS is still too shallow to give accurate simulation for each part of it, even only for the early vision. For a cascaded system, the error caused by incorrect modelling is likely to accumulate to generate disasters. Therefore, another thought is inspired to model each part of HVS and combine the results in parallel. This way is practical also because that modeling a stage of HVS is not necessarily requiring the responses of its prior stages. Generally, the parallel framework is more accurate and computationally faster, and the regression models produced by machine learning tools further evoke its development.

Secondly, HVS can be studied as an intact system. In other words, it can be treated as a black box. There are two ways to make use of the black box, both by finding the relationship between its outputs and inputs. The first approach is to compute the responses of it. HVS is capable to decompose visual signals according to different spatial frequencies, and its sensitivity to visual signals varies according to the frequencies. This decomposition can be modeled simply by frequency analysis. Because Fourier transform abandons the spatial information, other frequency analysis tools are generated, among which two most widely used are wavelet transform and short-time Fourier transform. Analyzing signals at different frequencies is also referred to as analysis at different scales. As the term implies, higher scale denotes lower frequency. So, the ability of frequency decomposition is also called the multi-scale property. Wavelet transform and short-time Fourier transform (STFT) are two commonly used tools to deal with the multi-scale property of HVS. Wavelet transforms adopt kernels to adaptively achieve the multi-scale functionality, while STFT adopts window functions for the same purpose in a manual manner, to some extent. Using different kernels or windows, the computed responses can be very different, so the selection among specific tools is a tricky task. Fortunately, studies in neurobiology have shown that using Gaussian window for STFT can generate responses that are very similar to the actual responses of HVS. STFT with Gaussian windows is called Gabor filter. Gabor filters can be set with different scales and orientations by setting different parameters, so that the multi-scale representation is constructed. With the decomposed visual signals, we can concentrate differently on responses at different frequencies. The sensitivity of HVS achieves maximum at a certain frequency, and drops with either rising or falling of the frequency, making HVS like a band-pass filter. This property is called contrast sensitivity function (CSF), where contrast can be regarded as another term to refer to the scale or frequency. CSF has received a lot of attention, and lots of mathematical models are constructed to simulate it. When analyzing the spatial frequencies, it is practical to refer to CSF to learn the favors of HVS. For instance, we can endow weights for different frequencies according to how much the specific frequency is appealing to HVS. Actually, most of modern IQA methods have taken the multi-scale property of HVS into account, because distortions may occur on various scales as well, and the multi-scale model can be effective in assisting the IQA methods, not to mention the ability of the frequency analyzing tools to extract quality-aware features, among which the applications of discrete cosine transforms, wavelet transforms and Gabor filters are extremely common. The second approach is based on information theory, to find out how the visual information is lost with distortions. Peak signal-to-noise ratio (PSNR) is based on the thought of information content. Problem of PSNR is the measured information is not necessarily related to human sensation upon image quality, so the definition of information is then modified, often referred to as "visual information", to suggest the relationship between the information and visual quality. In this sense, the definitions of signals and noises are also modified accordingly. Also, there are other ways of measuring the visual information for IQA, rather than the signal-to-noise ratio. The information fidelity of the distorted image and the mutual information shared by the distorted image and the reference are two widely known examples. Both the response-wise and the information-wise thoughts offer solutions to IQA in the matter that the inner structure of HVS is not exploited, so the solutions are mostly with good intuitiveness and simplicity.

Thirdly, many of the known properties of HVS have been concluded as abstract concepts and specific research fields have been set up. For two very important instances, visual saliency (VS) and just noticeable difference (JND) have been introduced in Sect. 6.4. These effects of HVS can be classified as neither the response of certain parts or cells of HVS nor the response of the whole system. This is fine because the purpose of IQA is neither to study all details of HVS as thoroughly as possible nor to treat HVS as a complete black box to only deem the interfaces as valuable. Reasonably, the abstract level should be moderate to avoid complex computations and to take care of HVS inner structures. Visual attention, including VS and JND, is a very good example of this moderate abstraction. VS or JND alone might not be a comprehensive description of the properties of HVS, but they are proved effective to be integrated with other feature extraction and pooling strategies to develop novel IQA methods. Moreover, because there are specific research fields of both VS and JND, many models are ready for us to employ, and the complication caused by studying lowest-level features is reduced. To conclude, it is promising to integrate our knowledge about HVS in neuroscience, psychology and other related fields to build more accurate and robust IQA methods.

References

Achanta R, Hemami S, Estrada F, Susstrunk S (2009) Frequency-tuned salient region detection. In: Proceedings of IEEE conference on computer vision and pattern recognition, Miami, FL, pp 1597–1604

Barlow HB (1957) Increment thresholds at low intensities considered as signal/noise discriminations. J Physiol 136(3):469–488

Bengio Y (2009) Learning deep hierarchies for AI. Found Trends Mach Learn 2(1):1–127

Boring EG (1944) Sensation and perception in the history of experimental psychology. Am J Psychol 57(1)

Borji A, Itti L (2013) State-of-the-art in visual attention modeling. IEEE Trans Pattern Anal Mach Intell 35(1):185–207

Borji A, Sihite DN, Itti L (2013) Quantitative analysis of human-model agreement in visual saliency modeling: a comparative study. IEEE Trans Image Process 22(1):55–69

Bouchakour M, Jeannic G, Autrusseau F (2008) JND mask adaptation for wavelet domain watermarking. In: International Conference on Multimedia and Expo, Hannover, Germany, pp 201–204

Bruce NDB, Tsotsos JK (2009) Saliency, attention, and visual search: an information theoretic approach. J Vis 9(3):1–24

Bruce NDB, Tsotsos JK (2005) An attentional framework for stereo vision. In: Proceedings of IEEE 2nd Canadian conference on computer and robot vision, Victoria, BC, Canada, pp 88–95

Chen M-J, Su C-C, Kwon D-K, Cormack LK, Bovik AC (2013) Full-reference quality assessment of stereopairs accounting for rivalry. Sig Process Image Commun 28:1143–1155

Chou CH, Li YC (1995) A perceptually tuned subband image coder based on the measure of just-noticeable-distortion profile. IEEE Trans Circuits Syst Video Technol 5(6):467–476

De Silva DVSX, FernandoWAC, Worrall ST, Yasakethu SL P et al (2010) Just noticeable difference in depth model for stereoscopic 3D displays. In: Proceedings of IEEE international conference on multimedia and expo, Singapore, Singapore, pp 1219–1224

De Silva DVSX, Ekmekcioglu E, Fernando WAC, Worrall ST (2011) Display dependent preprocessing of depth maps based on just noticeable depth difference modeling. IEEE J Sel Topics Signal Process 5(2):335–351

Ding Y, Zhao Y (2018) No-reference stereoscopic image quality assessment guided by visual hierarchical structure and binocular effects. Appl Opt 57(10):2610–2621

Du B, Yu M, Jiang G, Zhang Y, Shao F, et al (2016) Novel visibility threshold model for asymmetrically distorted stereoscopic images. In: Proceedings of 2016 visual communications and image processing, Chengdu, China, pp 1–4

Fan Y, Larabi M, Cheikh FA, Fernandez-Maloigne C (2017) Stereoscopic image quality assessment based on the binocular properties of the human visual system. In: IEEE international conference on acoustics, speech and signal processing, New Orleans, LA, pp 2037–2041

Fang Y, Wang J, Narwaria M, Callet PL, Lin W (2013) Saliency detection for stereoscopic images. In: Visual communications and image processing, Kuching, pp 1–6

Fang Y, Wang Z, Lin W, Fang Z (2014) Video saliency incorporating spatiotemporal cues and uncertainty weighting. IEEE Trans Image Process 23(9):3910–3921

Fezza SA, Larabi M, Faraoun KM (2014) Stereoscopic image quality metric based on local entropy and binocular just noticeable difference. In: IEEE international conference on image processing, Paris, France, pp2002–2006

Field DJ (1987) Relations between the statistics of natural images and the response properties of cortical cells. J Opt Soc Am 4(12):2379–2397

Geisler WS, Perry JS (1998) Real-time foveated multiresolution system for low-bandwidth video communication. Proceedings of SPIE—the international society for optical engineering, Ottawa, Canada, vol 3299, 294–305

Garcia-Diaz A, Fdez-Vidal XR, Pardo XM, Dosil R (2012) Saliency from hierarchical adaptation through decorrelation and variance normalization. Image Vis Comput 30(1):51–64

Geisler WS, Perry JS (1998) A real-time foveated multisolution system for low-band width video communication. In: Proceedings of SPIE—the international society for optical engineering, Ottawa, Canada, vol 3299, p 294

Gu Z, Ding Y, Deng R, Chen X, Krylov AS (2019) Multiple Just-Noticeable-Difference Based No-Reference Stereoscopic Image Quality Assessment. Appl Opt 58(2):340–352

Guo C, Zhang L (2010) A novel multiresolution spatiotemporal saliency detection model and its applications in image and video compression. IEEE Trans Image Process 19(1):185–198

Hachicha W, Beghdadi A, Cheikh, FA (2013) Stereo image quality assessment using a binocular just noticeable difference model. In: IEEE international conference on image processing, Melbourne, VIC, pp 113–117

References

Hahn PJ, Mathews VJ (1998). An analytical model of the perceptual threshold function for multichannel image compression. In: Proceedings of IEEE international conference on image processing, Chicago, IL, USA, USA, vol 3, pp 404–408

Harel J, Koch C, Perona P (2007) Graph-based visual saliency. In: Neural information processing systems, pp 545–552

Hou X, Zhang L (2007) Saliency detection: a spectral residual approach. In: IEEE conference on computer vision and pattern recognition, Minneapolis, MN, USA:2280–2287

Huang T-H, Liang C-K, Yeh S-L, Chen HH (2008) JND-based enhancedment of perceptibility for dim images. In: International conference on image processing, San Diego, CA, USA: 1752–1755

Hubel DH, Wiesel TN (1962) Receptive fields, binocular interaction and functional architecture in the cat's visual cortex. J Physiol 160(1):106–154

Hubel DH, Wiesel TN (1968) Receptive fields and functional architecture of monkey striate cortex. J Physiol 195(1):215–243

Itti L, Koch C, Niebur E (1998) A model of saliency-based visual attention for rapid scene analysis. IEEE Trans Pattern Anal Mach Intell 20(11):1254–1259

Jia Y, Lin W, Kassim AA (2006) Estimating just-noticeable distortion for video. IEEE Trans Circuits Syst Video Technol 16(7):820–829

Jiang Q, Duan F, Shao F (2014) 3D visual attention for stereoscopic image quality assessment. J Softw 9(7):1841–1847

Jones JP, Palmer LA (1987) An evaluation of the two-dimensional Gabor filter model of simple receptive fields in cat striate cortex. J Neurophysiol 58(6):1233–1258

Kandel ER, Schwartz JH, Jessel TM (2000) Principles of neural sciences. Mc-Graw-Hill

Koch C, Ullman S (1985) Shifts in selective visual attention: towards the underlying neural circuitry. Hum Neurobiol 4(4):219–227

Krüger N, Janssen P, Kalkan S, Lappe M, Leonardis A, Piater J, Rodríguez-Sánchez AJ, Wiskott L (2013) IEEE Trans Pattern Anal Mach Intell 35(8):1847–1871

Le Meur O, Le Callet P, Barba D, Thoreau D (2006) A coherent computational approach to model bottom-up visual attention. IEEE Trans Pattern Anal Mach Intell 28(5):802–817

Legras R, Chanteau N, Charman WN (2004) Assessment of just-noticeable differences for refractive errors and spherical aberration using visual simulation. Optom Vis Sci 81(9):718–728

Li X, Wang Y, Zhao D, Jiang T, Zhang N (2011) Joint just noticeable difference model based on depth perception for stereoscopic images. In: Proceedings of IEEE international conference on vision communication and image processing, pp 1–4

Li Y (2019) No-reference stereoscopic image quality assessment based on visual attention and perception. IEEE Access 7:46706–46716

Liu A, Lin W, Paul M, Deng C, Zhang F (2010) Just noticeable difference for images with decomposition model for separating edge and textured regions. IEEE Trans Circuits Syst Video Technol 20(11):1648–1652

Liu Y (2016) Stereoscopic image quality assessment method based on binocular combination saliency model. Sig Process 125:237–248

Marr D (1983) Vision: a computational investigation into the human representation and processing of visual information. W.H. Freeman and Company, New York, NY

Murala S, Maheshwari RP, Balasubramanian R (2012) Local tetra pattern: a new feature descriptor for content-based image retrieval. IEEE Trans Image Process 21:2874–2886

Nguyen A, Kim J, Oh H, Lin W, Lee S (2019) Deep visual saliency on stereoscopic images. IEEE Trans Image Process 28(4):1939–1953

Ojala T, Pietikäinen M, Mäenpää T (2002) Multiresolution gray-scale and rotation invariant texture classification with local binary pattern. IEEE Trans Pattern Anal Mach Intell 24(7):971–987

Orban GA (2008) Higher order visual processing in macaque extrastriate cortex. Physiol Rev 88:59–89

Peterson HA, Peng H, Morgan JH, Pennebaker WB (1991) Quantization of color image components in the DCT domain. International society for optics and photonics, pp 210–222

Qi F, Jiang T, Fan X, Ma S, Zhao D (2013) Stereoscopic video quality assessment based on stereo just-noticeable difference model. In: Proceedings of 20th IEEE international conference on image processing, pp 34–38

Qi F, Zhao D, Fan X, Jiang T (2013) Stereoscopic video quality assessment based on stereo just-noticeable difference model. In: International conference on image processing, Melbourne, Australia, pp 34–38

Qi F, Zhao D, Gao W (2015) Reduced reference stereoscopic image quality assessment based on binocular perceptual information. In: IEEE Transactions on Multimedia 17(12):2338–2344. https://doi.org/10.1109/TMM.2015.2493781

Safranek RJ, Johnston JD (1989) A perceptually tuned sub-band image coder with image dependence quantization and post-quantization data compression. In: Proceedings of IEEE conference on acoustic, speech, and signal processing, pp 1945–1948

Shao F, Lin W, Gu S, Jiang G, Srikanthan T (2013) Perceptual full-reference quality assessment of stereoscopic images by considering binocular visual characteristics. IEEE Trans Image Process 22(5):1940–1953

Seo HJ, Milanfar P (2009) Static and space-time visual saliency detection by self-resemblance. J Vis 9(12):15

Tenenbaum FE, David SV, Singh NC, Hsu A, Vinje WE, et al (2001) Estimating spatio-temporal receptive fields of auditory and visual neurons from their responses to natural stimuli. Netw: Comput Neural Syst 12(3):289–316

Toprak S, Yalman Y (2017) A new full-reference image quality metric based on just noticeable difference. Computer Standards & Interfaces 50:18–25

Walther D, Koch C (2006) Modeling attention to salient proto-objects. Neural Netw 19(9):1395–1407

Wang X, Ma L, Kwong S, Zhou Y (2018) Quaternion representation based visual saliency for stereoscopic image quality assessment. Sig Process 145:202–213

Watson AB, Yang GY, Solomon JA, Villasenor J (1997) Visibility of wavelet quantization noise. IEEE Trans Image Process 6(8):1164–1175

Wei Z, Ngan KN (2009) Spatio-temporal just noticeable distortion profile for grey scale image/video in DCT domain. IEEE Trans Circuits Syst Video Technol 19(3):337–346

Xu X, Zhao Y, Ding Y (2017) No-reference stereoscopic image quality assessment based on saliency-guided binocular feature consolidation. Electron Lett 53(22):1468–1470

Yang J (2016) Quality assessment metric of stereo images considering cyclopean integration and visual saliency. Inf Sci 373:251–268

Yang J, Ji C, Jiang B, Lu W, Meng Q (2018) No reference quality assessment of stereo video based on saliency and sparsity. IEEE Trans Broadcast 64(2):341–353

Yang XK, Lin W, Lu ZK, Ong EP, Yao SS (2005) Just noticeable distortion model and its applications in video coding. Sig Process Image Commun 20:662–680

Yao Y, Shen L, Geng X, An P (2017) Combining visual saliency and binocular energy for stereoscopic image quality assessment. Springer 685:104–114

Zhang L, Gu Z, Li H (2013) SDSP: a novel saliency detection method by combining simple priors. In: Proceedings of IEEE international conference on image processing, Sanya, Hainan Island, China, pp 171–175

Zhang L, Shen Y, Li H (2014) VSI: a visual saliency-induced index for perceptual image quality assessment. IEEE Trans Image Process 23(10):4270–4281

Zhang W, Luo T, Jiang G, Jiang Q, Ying H, et al (2016) Using saliency-weighted disparity statistics for objective visual comfort assessment of stereoscopic images. 3DR Express 7:17

Zhang Y, Jiang G, Yu M, Chen K (2010) Stereoscopic visual attention model for 3D video. In: Proceedings of 16th international conference on advanced multimedia modelling, Chongqing, China, pp 314–324

References

Zhao Y, Chen Z, Zhu C, Tan Y-P, Yu L (2011) Binocular just-noticeable difference model for stereoscopic images. IEEE Signal Process Lett 18(1):19–22

Zhong R, Hu R, Shi Y, Wang Z, Han Z, et al (2012) Just noticeable difference for 3D images with depth saliency. In: Proceedings of Pacific-Rim conference on multimedia, Berlin, Germany, pp 414–423

Zhong R, Hu R, Wang Z, Wang S (2015) 3D hybrid just noticeable distortion modeling for depth image-based rendering. Multim Tools Appl 74(23):10457–10478

Zhou L (2012) A new just-noticeable-distortion model combined with the depth information and its application in multi-view video coding. In: Proceedings of 8th international conference on intelligent information hiding & multimedia signal processing, Guanajuato, Mexico, pp 246–251

Chapter 7
Stereoscopic Image Quality Assessment Based on Deep Convolutional Neural Models

Abstract The deep convolutional neural network (CNN) has achieved great success in image process areas in recent years. Many image quality assessment methods directly use CNN for quality prediction. Optimizing deep convolutional neural network with high generalization ability needs a huge amount of data, however, the most popular IQA databases are usually too small. Therefore, transfer learning and patch-wise strategy are developed to realize data enhancement. On the basis of alleviating the insufficient training data, some methods improve the CNN framework to better simulate HVS, and the implementation details are described in this chapter. Finally, some necessary related knowledges about CNN-based IQA methods are introduced.

Keywords Convolutional neural network · Stereoscopic image quality assessment · Transfer learning · Patch-wise · Saliency-guided

7.1 Introduction

Recently, convolutional neural networks (CNNs) have shown standout performance on many computer vision and image processing applications (e.g., image classification, image recognition and semantic segmentation) (Krizhevsky et al. 2012; Dou et al. 2017; Shelhamer et al. 2017). In contrast to traditional image processing approaches, CNNs can automatically learn tasks-related visual features, simulate the procedure of human perceiving and deal with image information.

In early days, the principle of CNN was first introduced in LeCun et al. (1998). The deep model, called LeNet-5, was designed for handwritten digit recognition, consisting of a multi-layer artificial neural network. The detailed network architecture adding convolutional layers is presented in Fig. 7.1. The introduction of convolutional layers made the deep models more effectively learn feature representations of the neuron's receptive field of the image inputs due to the advantages (i.e., sparse interactions, parameter sharing and equivariant representations). Yet, limited by inadequate training data and poor computing power, LeNet-5 cannot perform as well as the handwritten task on more complex image and video classification tasks.

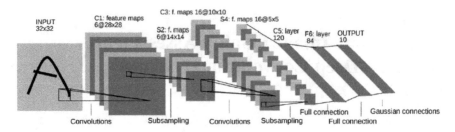

Fig. 7.1 The architecture of LeNet-5 for handwritten digit classification tasks (LeCun et al. 1998)

With the great improvements in the power of graphics processor units, the difficulties involved in the step of deep CNN model training have been overcome gradually. More and more complex CNNs have been designed for various visual tasks, which can be trained and tested effectively and rapidly, achieving state-of-the-art performance. In 2012, Krizhevsky et al. (2012) proposed a classical CNN architecture for image classification, namely AlexNet, which won the title in ImageNet competition by far surpassing the second place. Henceforth, CNN and even deep learning began to attract extensive attention.

Next, a popular tendency is that CNNs are getting deeper and more complex. For example, VGGNet (Simonyan and Zisserman 2015) directly pushed the network's depth into 19 layers with over one hundred million parameters, which is critical for achieving good results. GoogLeNet, proposed in Szegedy et al. (2015), adopted the idea that deeper and wider networks can easier to achieve promising prediction performance than that using shallower architectures. However, it would also dramatically increase computational requirement of hardware resources. Most notably, the deeper architecture also increased the complexity of the network, which made it hard to optimize, and even had a tendency of overfitting.

To solve the puzzled problem, various research has attempted to optimize and improve the architecture of CNN in various aspects, in which ResNet (He et al. 2016) presented a residual learning framework to ease the training of networks and easier to optimize than those used previously. By increasing the depth up to 152 layers, ResNet could better learn visual feature representations, and meanwhile the introduction of residual learning also decreased the computing complexity during training, which made the network win the 1st place on the ILSVRC 2015 classification task.

In recent years, a number of models have been proposed for various computer vision tasks with significant improvements than before, delivering the message that it is CNNs that can approximate the target function step by step to get better feature representations (Chan et al. 2015; Zhang et al. 2015). In particular, most of top-ranked models in the current computer vision competitions (e.g., image recognition and image classification) adopt CNN architectures. Thus, motivated by the great success of CNN-based image processing applications, how to apply the CNN into IQA tasks has become the focus of research.

7.2 Stereoscopic Image Quality Assessment Based on Machine Learning

Actually, machine learning has played a significant role in IQA fields over the past three or four years, especially for 2D IQA tasks. For example, shallow mapping engines such as support vector regression or random forests have been applied to obtain good results in quality prediction (Ding et al. 2018; Sun et al. 2018; Ye et al. 2012). These methods aimed to extract hand-craft visual features from the synthesize cyclopean images or monocular views of stereopairs, which could be aggregated into the final objective scores using shallow regressors. Afterwards, since the deep features from shallow CNNs can automatically capture more useful visual information related to image quality and human perceptions rather than extracting hand-crafted features, the straightforward way was that tasks-related discriminative features were learned using the CNN over a number of training iterations on a large dataset from high-dimensional raw images. Then the learned feature representations were subsequently aggregated into the global score by shallow regressors (Lv et al. 2016; Yang et al. 2019). Recently, end-to-end neural networks are expected to obtain a quality-aware CNN architecture, from which we can directly generate the predicted objective scores of distorted images by an effective end-to-end optimization (Li et al. 2016). In the trained CNNs, the learned feature maps contain rich and abstract information sensitive to image perceptual quality, which are regressed into the objective score by the following fully connected (FC) layers. For example, Lv et al. (2016) extracted a set of visual features to synthesize the predicted quality rating, where manual features were obtained by simulating binocular rivalry and suppression, and meanwhile the deep features were generated from the trained CNN architecture. Another example of CNN-based SIQA is proposed by Zhang et al. (2016), which utilizes CNN to learn the local structure of distorted images, and then employs the multilayer perceptron to pool the learned local structure to the final quality score of the stereoscopic images. Figure 7.2 displays the four different general frameworks of IQA fields that have been introduced before.

With the increasement of the complexity and depth of neural networks, the application of CNNs to IQA tasks has faced a significant obstacle, which is a lack of adequate training data assigned to corresponding ground-truth human subjective scores. Although in principal CNNs can achieve greatly promising performance in quality prediction, there are some negative effects (e.g., overfitting and non-convergence) in directly optimizing CNN-based IQA models on a small IQA database. Currently available existing 2D IQA databases such as LIVE (Sheikh et al. 2006) and TID2013 (Ponomarenko et al. 2015), and 3D IQA databases including LIVE 3D Phase I and II (Moorthy et al. 2013; Chen et al. 2013) are far from sufficient to train the complex CNN models. The basic information about image numbers of mainly public 2D and 3D IQA databases are shown in Table 7.1. Taking the LIVE 3D Phase I and II databases as examples, there are only total contain 725 distorted stereopair samples, far smaller than the ImageNet dataset with over 50 million labeled training data for image classification. In addition, creating larger subjective 3D IQA

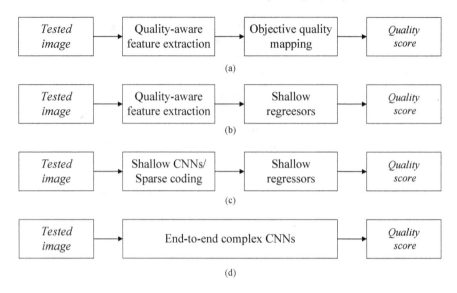

Fig. 7.2 Four different frameworks for IQA tasks. **a** Traditional IQA architectures, **b** shallow regressors, **c** shallow CNNs for feature extraction and **d** end-to-end CNNs

Table 7.1 Basic information about the public 2D and 3D IQA databases

Tasks	Databases	Total images	Resolution	Label range
2D IQA	LIVE	779	Various	[0, 100]
	CSIQ	866	512 × 512	[0, 1]
	TID2008	1700	384 × 512	[0, 9]
	TID2013	3000	384 × 512	[0, 9]
3D IQA	LIVE 3D I	365	360 × 640	[−10, 70]
	LIVE 3D II	360	360 × 640	[0, 80]
	WATERLOO II	460	1080 × 1920	[0, 100]
Image classification	ImageNet	>50 million	Various	N/A

datasets is a formidable problem because reliable subjective labels are not easily obtained. In general, collecting human subjective judgments in controlled laboratory like (Sheikh et al. 2006; Ponomarenko et al. 2015) is a time-consuming procedure and even out of question.

7.3 Stereoscopic Image Quality Assessment Based on Transfer Learning

7.3.1 Theoretical Basis for Transfer Learning

As described before, the excellent performance of CNN-based IQA models depends on a large amount of prior data while the biggest obstacle in IQA tasks is actually lack of training data. The early thought was directly borrowing some relatively stable pre-trained CNN models from another relevant tasks. Considering FR image prediction problem as an example, Gao et al. (2017) extracted deep feature maps from the reference images and its distorted versions by feeding them into the CNN model that had been pre-trained on the ImageNet. Local similarities between the extracted feature maps from reference and distorted images were then computed and pooled to arrive at global subjective scores. Note that, the CNN model was not fine-tuned on any 2D IQA datasets during the procedure of feature extraction. Nevertheless, recent studies indicated that the deep features directly extracted from the pre-trained model designed for other targets (e.g., object detection or image classification) are not ensured to sensitive to image corruption (Bianco et al. 2018). Further, to solve the critical problem about how to transform such pre-trained CNN features into quality-aware features from little labeled training datasets, transfer learning is proved to be an effective strategy, where the complex CNN model pre-trained on a specific large image database is fine-tuned on little IQA databases to fit the quality prediction task of images, as depicted in Fig. 7.3.

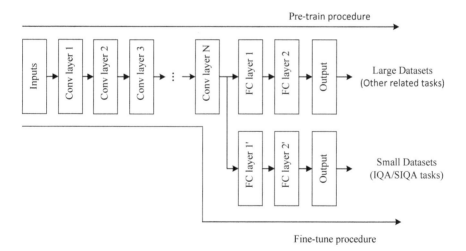

Fig. 7.3 The general framework of transfer learning. In the procedure of pre-training, the CNN is trained sufficiently and effectively using image samples from a large dataset. Then, the fine-tuning step is implemented by substituting several specific network layers (i.e., FC layer 1 and FC layer 2) with several new network layers (i.e., FC layer $1^{'}$ and FC layer $2^{'}$) to fit the specific training targets

To the best of our knowledge, transfer learning has been applied in several small sample tasks and achieved significant progress (Pan and Yang 2009; Bengio 2011). Before employing the strategy of transfer learning into 2D or 3D IQA tasks, brief analysis is required in this section. In general, unless the targets of our research are very different from the pre-trained image processing fields, transfer learning can always be utilized to fine-tune the pre-trained CNN model to avoid the need of training from scratch. Experiments have observed that deep features trained from large natural images contain some more universal and general visual information that can be applied to other related tasks, like curves and edges, unless the two proposed tasks are significantly different. When fine-tuning the pre-trained CNN model, the universal features can be transformed into more specific feature vectors related to our targets from the trainable later layers. Therefore, it helps to demonstrate the feasibility of transfer learning applications in 2D and 3D IQA.

In addition, when only fine-tuning the later layers of the pre-trained CNN model instead of all trainable layers, little labeled data are sufficient for generating tasks-related visual features, which is very suitable for the specific IQA tasks. With common data augmentation approaches, the prediction performance can be further improved via transfer learning.

7.3.2 From Image Classification to Quality Regression Task

Based on the brief analysis, transfer learning is expected to solve the large-scale training data dependence existing in current CNN-based IQA models, which was first applied in the field of 2D IQA tasks. Because the targets of 2D IQA is to predict a global quality score of the distorted image, which is naturally considered as a regression problem. The regression problem means that a specific scalar is recognized as the ground-truth label during training, such as DMOS = 50.34, DMOS = 10.28, from which these CNN models can be utilized to predict image quality scores. Henceforth, most deep learning-based IQA methods adopted the idea to realize the goal of quality prediction in an end-to-end optimization, in which deep models were fine-tuned to learn quality-related feature representations from image classification to quality regression tasks naturally. For example, Li et al. (2016) designed a network in network model pre-trained on ImageNet database to enhance the abstraction ability of CNN model. And the final layer of the modified CNN model was replaced by the regression layer which aggregated the learned features into subjective scores by being fine-tuned on specific 2D IQA datasets. Of course, even though transfer learning can make up for the problem of lack of training data to some extent, most later studies would like to combine data augmentation approaches to address the problem to a greater extent, expecting to achieve better results (Zhang et al. 2016; Sun et al. 2020; Zhou et al. 2019; Liu et al. 2017; Dendi et al. 2019). The main data augmentation approaches applied into IQA tasks could be divided into two main parts: patch-wise (Zhang et al. 2016; Sun et al. 2020; Zhou et al. 2019)

7.3 Stereoscopic Image Quality Assessment Based on Transfer Learning

and more training data creation (Liu et al. 2017; Dendi et al. 2019), which will be discussed in detail in the following sections.

7.3.3 From Image Classification to Quality Classification Task

The fact has been declared that the more similar the two tasks are, the more likely it is to achieve satisfactory results from transfer learning. This forced the generation of the idea of fine tuning the task of quality prediction as a classification problem. On the one hand, recognizing IQA as a classification problem, it could increase the relevance between our tasks and image classification tasks. On the other hand, it could make up for the shortage of training samples in each category as much as possible. For 2D IQA tasks, Hou et al. (2015) first transferred quality prediction problem to quality classification problem by classifying distorted images to five quality levels according to corresponding subjective scores. The lower level features of distorted images were extracted and fed into the deep belief network (DBN) for layer-by-layer pre-training. Then, the DBN was fine-tuned to classify image grades by maximizing the probabilistic distribution. This work was improved in Hou and Gao (2015) by introducing saliency maps to improve the accuracy of quality prediction. The first attempt for SIQA tasks we would like to introduce is Ding et al. (2018), in which the pre-trained CaffeNet architecture (Jia et al. 2014) was fine-tuned in 2D and 3D IQA datasets to adaptively extract quality-aware features as a quality classification task, and the adaptive features were aggregated into a final objective score. The framework of this paper is illustrated in Fig. 7.4.

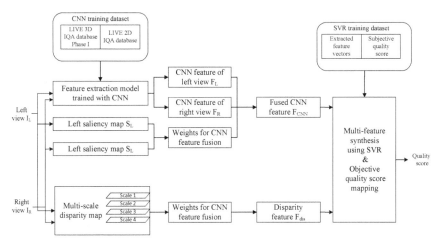

Fig. 7.4 The framework proposed in Ding et al. (2018)

In general, CNNs are composed of a set of stacked convolutional layers that are recognized as good feature generators corresponding to specific image problems. A reasonable assumption is that a pre-trained CNN model in the image classification task can classify image contents and details well. Thus, when the pre-trained CNN is fine-tuned in a similar classification task, it is more likely to achieve promising results than other irrelevant fields. Based on this assumption, the pre-trained CNN model is fine-tuned from scratch to classify distorted images according to corresponding image quality, from which adaptive quality-aware monocular features can extracted from the high-level layers of the CNN. The main steps to turn the regression task into the classification task are as follows:

1. The subjective quality ratings of distorted images are ranked and grouped into 6 equal-size classes with corresponding categories from 0 to 5 (i.e., 0 represents highest quality images and 5 represents lowest quality images)
2. The corresponding distorted image are also classified into 6 groups according to their corresponding subjective quality scores.

To make the pre-trained CNN model progressively more specific and perceptive to SIQA tasks, the authors redesigned the last three FC layers to fit the classification task with six image quality categories, similarly displayed in Fig. 7.3. Therefore, the CaffeNet model can be trained efficiently by fixing the weighting parameters of the convolutional layers already pre-trained on ImageNet and only training the new FC layers, in which quality-aware monocular features of left or right view of a distorted stereopair can be obtained from the penultimate FC layer. In the meanwhile, visual saliency, as described in Sect. 6.1, is utilized to combine the monocular features between left view and right view of a stereopair (Li et al. 2015). Let F_L and F_R denote the extracted deep monocular features of left view and right views in a distorted stereopair, S_L and S_R represent the saliency maps of left and right monocular image, respectively. The final binocular features can be calculated by the following function:

$$F_{CNN} = F_L W_L + F_R W_R \quad (7.1)$$

where W_L and W_R are denoted by

$$W_L = \frac{\sum_{i=1}^{N} S_L(i)}{\sum_{i=1}^{N} S_L(i) + \sum_{i=1}^{N} S_R(i)}$$
$$W_R = \frac{\sum_{i=1}^{N} S_R(i)}{\sum_{i=1}^{N} S_L(i) + \sum_{i=1}^{N} S_R(i)} \quad (7.2)$$

where N is the number of image pixels of a saliency map.

Besides, as the complementation for CNN features, multi-scale statistical features are extracted from binocular disparity maps of the stereoscopic image, which is described as below.

7.3 Stereoscopic Image Quality Assessment Based on Transfer Learning

$$F_{i,D1} = \frac{M_i}{M_i + K_i + S_i}$$
$$F_{i,D2} = \frac{K_i}{M_i + K_i + S_i}$$
$$F_{i,D3} = \frac{S_i}{M_i + K_i + S_i} \quad (7.3)$$

where M, K and S represent mean value, kurtosis and skew of disparity maps, respectively. i is the scale of the disparity map, in which i is range from 1 to 4.

After the extracted CNN features and the statistical features of disparity maps are obtained, SVR is adopted to construct a regression function that can aggregate these multiple features into the final objective quality score.

The improvement version of Ding et al. (2018) is proposed in Xu et al. (2019), where the authors analyzed the role of transfer learning in SIQA tasks systematically. In this paper, conclusion has been given that fine-tuning the last FC layer for image quality classification is enough to achieve promising performance for NR SIQA. Especially, two classical CNN models with different structure and depth, CaffeNet and GoogLeNet (Szegedy et al. 2015), were selected to support the generality of conclusion. Experiments proved that the target function of both the two pre-trained CNN models will converge in the fine-tuning stage gradually, which further delivers the message that the application of fine-tuning strategy can obtain pretty good performance for predicting stereopairs' quality regardless of the diversity in the structure and depth of networks.

7.4 Stereoscopic Image Quality Assessment Based on Patch-wise Models

7.4.1 Theoretical Basis for Patch-wise Strategy

However, there still some concerns about the effects of transfer learning on SIQA fields. Although the research targets of SIQA can be transferred into the quality classification problem to cater to the pre-trained deep models for image classification tasks, it is still not confirmed whether or not the features that are learned from the pre-training process are sensitive to the degree of corruption owing to different purposes of feature learning in the two training stages (Kim and Lee 2017). In addition, some researchers still hope to address the problem of insufficient data at the root instead of only relying on transfer learning. Therefore, another common strategy, called patch-wise, was adopted into the tasks related to quality prediction to alleviate the obstacle of inadequate training data. The patch-wise strategy can be realized in general by splitting the image samples into numerous small fixed-size patches. As a kind of supervised learning, the training procedure of neural networks must be marked with the corresponding ground-truth labels, which can optimize the network toward the training target. To begin with, according to the assumption that every training stereoscopic image had homogeneous distortion degree in available IQA datasets, many researchers trained the CNNs by assigning the subjective score to each patch of a distorted image as the ground-truth annotation, which is depicted in Fig. 7.5. The general framework with the patch-wise strategy can be divided into two steps as follows:

Training step 1: For IQA tasks, the distorted images are split into several fixed-size image patches, and meanwhile each image patch is assigned a global score the same as the one of its corresponding original image. Local quality-aware CNN model is trained by the training data pair with fixed-size image patches as training images and the allocated subjective score as ground-truth label.

Training step 2: The local deep features extracted from the shared local CNN model are aggregated into the global objective score in an end-to-end manner.

As a matter of fact, the IQA based on the patch-wise strategy alleviates the problem about inadequate training data to some extent, and even leads to a significant performance improvement. Thus, many studies have accepted the patch-wise strategy and applied them in predicting quality of images.

To the best of our knowledge, the framework proposed in Kang et al. (2014) is the first to apply the strategy of patch-wise to the field of IQA, wherein a shallow CNN model was designed to extract discriminative features from a three-dimensional input image patch. Each image patch was assigned the same subjective quality score as the proxy label during training. Following the strategy of patch-wise, the predicted scores of image patches were then averaged to obtain the final global score.

However, the assumption is proved to be incorrect in later studies, and the objective score value of each image patch from the stereopair is not necessarily the same as the overall label assessed from the corresponding stereopair. In order to get more reliable ground-truth labels for each distorted stereo image patch, several studies have shown that local quality scores calculated by some traditional SIQA methods such as SSIM, GSIM and GMD can be recognized the proxy ground truth quality

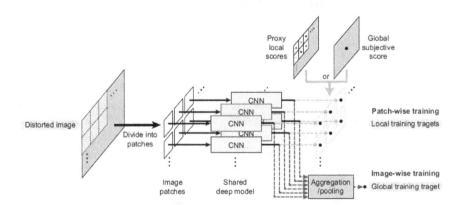

Fig. 7.5 The general framework of patch-wise model in early years

labels. Thus, the SIQA based on patch-wise models can be improved and redefined as two separate stages: a pre-training local feature learning stage using image patches and corresponding algorithm-generated labels, followed by a stage of global score regression, as also shown in Fig. 7.5. For example, Kim and Lee (2017) designed a local patch-based CNN model to predict the quality of plane images, whereby local quality annotations were generated as the proxy ground-truth label by the obtained FR SIQA algorithms.

7.4.2 Patch-wise Strategy with Global Subjective Score

The first application of the patch-wise thought in SIQA tasks was proposed in Zhang et al. (2016), where the authors used CNN architecture to learn local feature representations of stereopair patches in left, right and difference images. In details, the paper proposed two deep models according different inputs, namely one-column CNN with difference image patches as the input, and three-column CNN with stereopair image patches from left, right and difference images as CNN inputs. Here we only want to introduce three-column CNN architecture, as shown in Fig. 7.6.

Different with 2D nature images, we have known that 3D images contain more visual information such as difference images. Thus, difference images are considered into the framework as CNN inputs. Generally, the difference images I_D can be obtained from left images I_L and right images I_R in stereopairs:

$$I_D(x, y) = I_L(x, y) - I_R(x, y) \tag{7.4}$$

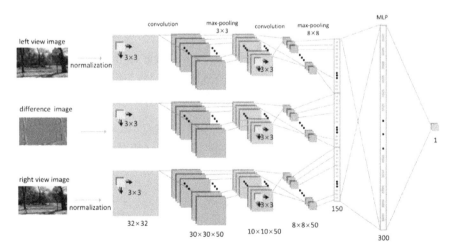

Fig. 7.6 The framework of three-column CNN model proposed in Zhang et al. (2016)

As aforementioned, stereoscopic images are split into several 32 × 32 image patches to expand training data from left, right and obtained difference images, respectively. With image patches from left, right and difference images as inputs, three CNN branches are designed, where each CNN branch contains two layers of convolution and max-pooling layers. The three CNN branches are utilized to learn local structures of left, right and difference images, respectively. Then the multilayers perception with two FC layers are used to summarize the feature representations and generate the final predicted score.

7.4.3 Patch-wise Strategy with Generated Quality Map

Different from the subjective score as training label for each patch, Oh et al. (2017) designed a local patch-wise CNN to learn local feature structures of the stereopair, where the corresponding ground-truth label of each image patch was assigned from SSIM metric. Figure 7.7 illustrates the deep CNN model, mainly dividing into two training steps: local structure feature learning and global score regression.

Here we mainly focus on how the local quality score for each image patch is formed instead of the training procedure. That's because how to train a quality-aware CNN model has been introduced before. To train the local structures from the generated image patches, the ground-truth label for each image patch need to be assigned, where the authors designed a traditional SIQA method to solve the obstacle. The details of the framework can be separated into three steps as follows:

Step 1: The cyclopean image I_C can be modeled from the left image I_L and right image I_R by the following function at pixel (x, y).

$$I_C(x, y) = W_L(x, y) \times I_L(x, y) + W_R(x+d, y) \times I_R(x+d, y) \tag{7.5}$$

where d represents pixel different value between I_L and I_R in coordinates (x, y). W_L and W_R are the normalized weights calculated by the Gabor filter, as already described in Sect. 5.2.

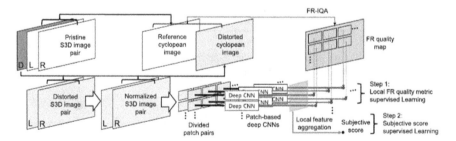

Fig. 7.7 The framework proposed in Oh et al. (2017), which can be divided into two stages: training step 1 with local structure feature learning and training step 2 with global score regression

7.4 Stereoscopic Image Quality Assessment Based on Patch-wise Models 147

Fig. 7.8 Examples of obtaining the local quality score: **a** and **b** the left and right views of a stereopair, **c** and **d** the synthesized cyclopean reference and distorted images, **e** local quality map calculated from cyclopean SSIM metric, **f** the ground-truth labels for each patch, where brighter regions represent higher quality of corresponding image patches

Step 2: Some classical 2D IQA can be applied into the synthesized cyclopean reference and distorted images, among which SSIM was finally selected to generate local quality maps due to its reasonable performance in predicting plane image quality.

Step 3: After the local quality maps are derived, the ground-truth local quality label l_n is obtained by averaging the local quality map M_{SSIM} corresponding to each image patch over the 18×16 window:

$$l_n = \frac{1}{18 \times 16} \sum_{x,y \in p_{cn}} M_{SSIM}(x, y) \qquad (7.6)$$

where $P_c = \{p_{c1}, p_{c2}, ..., p_{cN}\}$ represent the image patch set from the cyclopean distorted image.

Figure 7.8 depicts the example in LIVE 3D Database Phase I for obtaining the ground-truth local quality label, where (a) and (b) are the left and right views from a pristine stereopair, (c) and (d) represent the synthesized cyclopean reference and distorted images from (a) and (b). The generated local quality map and its corresponding averaging version are shown in Fig. 7.8e, f.

7.4.4 Saliency-guided Local Feature Selection

On the basis of alleviating the insufficient training data by adopting the patch-wise strategy, some studies began to shift the focus to improve the CNN framework to better simulate HVS in human brain. The example introduced in Sun et al. (2020) utilized visual saliency to guide deep visual feature extraction in the stage of global score regression, and meanwhile together the patch-wise strategy and three-column CNN architecture into SIQA tasks, as depicted in Fig. 7.9.

According to different inputs, the authors also proposed two CNN architectures, where the model we would like to introduce is the three-column CNN model. Firstly, motivated by binocular and monocular properties in human brain, this article designed three CNN stream models, in which two CNN branches are trained for simulating monocular mechanism with the left and right views as inputs, and the rest one CNN branch is developed for binocular visual property with the synthesized

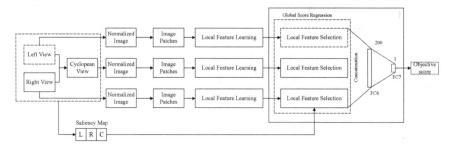

Fig. 7.9 Framework of three-column CNN model proposed in Sun et al. (2020), which consists of two stages: local quality feature learning and global score regression

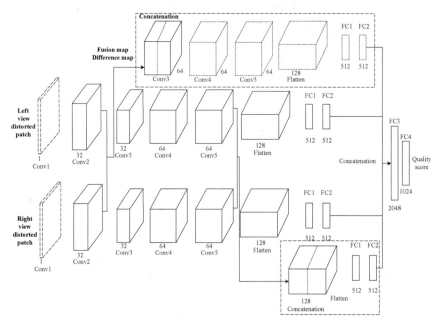

Fig. 7.10 Framework of the dual-stream interactive CNN model proposed in Zhou et al. (2019) containing four subnetworks: two networks for the left and right views, and the rest for the fusion and difference maps

7.4 Stereoscopic Image Quality Assessment Based on Patch-wise Models

cyclopean image as the input. In order to obtain more training data, the patch-wise strategy is also adopted with the generated local quality map as the ground-truth label assigned for each patch. In addition, experiments have proved that different regions in the stereopair have different contributions on aggregating the global subjective rating. In general, the image patches on salient regions attract more visual attention than that on non-salient regions, and even those patches on homogeneous regions could introduce large prediction bias (Cheng et al. 2017; Jiang et al. 2014). Thus, the authors developed a strategy of saliency-guided feature selection in the stage of global score regression, which means that salient and non-salient regions can be distinguished by comparing the given saliency threshold with average salient value of corresponding saliency map generated from the distorted image. For generating the global objective score (Q_O), only deep features obtained from monocular and binocular image patches in salient regions are trained and regressed into the global objective score in an end-to-end manner, aiming to eliminate negative effects of non-salient image patches. The procedure can also be expressed in the following function.

$$\left[Q_{j=1,2,\ldots,K}\right]_v = \begin{cases} q_{vm}, & \text{if } S_v(p_m) > T \\ skipped, & \text{otherwise} \end{cases}, \quad m \in [1, N_p], v \in \{C, L, R\}$$

$$\left[Q_{j=1,2,\ldots,3K}\right] = \text{Concat}\left(\left[Q_{j=1,2,\ldots,K}\right]_L, \left[Q_{j=1,2,\ldots,K}\right]_R, \left[Q_{j=1,2,\ldots,K}\right]_C\right)$$

$$Q_O = f_\theta\left(\left[Q_{j=1,2,\ldots,3K}\right]\right) \tag{7.7}$$

where q_{vm} represents the trained deep visual features extracted from the image patches p_{vm}, in which $v \in \{C, L, R\}$ denotes the cyclopean, left or right view. S_v is the corresponding averaged saliency map of the cyclopean or monocular view. Concat() is the feature combination function in deep learning. $f_\theta()$ represents the regression procedure with weighting parameters θ. T denotes the specific saliency threshold, and N_p is the total number of non-overlapping image patches from a stereopair.

7.4.5 Dual-stream Interactive Networks

The last example we would like to discuss is a dual-stream interactive CNN model proposed in Zhou et al. (2019) shown in Fig. 10. The authors first designed a two-stream CNN network for the left and right views, where each subnetwork shares the same CNN model, consisting of five convolutional layers and two FC layers. The detail network parameters can be found in Zhou et al. (2019). Inspired by the dual-stream interaction mechanism of human visual cortex responses, the multilayers network interaction between left and right view subnetwork was then developed to expect to obtain binocular interaction properties. Note that, the authors also adopted the patch-wise strategy, where multiple patch pairs sampled from distorted stereopairs were fed into the dual-stream CNN model in the training procedure. Through the deeper analysis of the relationship between deep neural networks and hierarchical human visual system, the second convolutional layer (*Conv*2) and the fifth convolutional layer (*Conv*5), representing the lower and higher visual features, were exploited to generate two concatenation subnetworks, respectively. Let F_L and F_R represent the convolutional feature maps generated from *Conv*2 and *Conv*5. The fusion and difference maps can be calculated by the summation and subtraction operations from F_L and F_R as follows:

$$S^+ = F_L + F_R$$
$$S^- = F_L - F_R \qquad (7.8)$$

where S^+ and S^- are the generated fusion and difference maps, which were concatenated and fed into the concatenation subnetwork.

To more intuitively observe the effects of summation and subtraction operations, examples of distorted stereopairs with different distortion types and levels and its corresponding fusion as well as difference maps are shown in Fig. 7.11.

Finally, there are total four subnetworks in this paper. For generating the predicted quality score, these feature vectors generated from the four subnetworks were concatenated and aggregated by the FC layers as follows:

$$Q_O = f_\theta(\text{Concat}(V_L, V_R, V_{Conv2}, V_{Conv5})) \qquad (7.9)$$

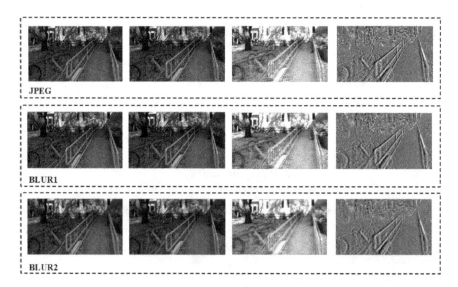

Fig. 7.11 Examples of the fusion and difference maps generated from distorted stereopairs. The first and second columns represent the left and right views of different distortion types/levels. The third column is the corresponding fusion maps generated from left and right distorted images by summation operation. The last column denotes the difference maps calculated by subtraction operation between the left and right views

where V_L, V_R, V_{Conv2} and V_{Conv5} denote the output feature vectors generated from the four subnetworks, respectively. $f_\theta()$ represents the regression procedure of FC layers.

7.5 New Tendency for Exploiting CNN-Based SIQA Tasks

We have noticed that inadequate training data limits the structure and training performance of CNNs in the previous sections. A straightforward insight is to create more distorted image samples for better representations of CNN models. Although the collection of subjective score for each distorted image is extremely difficult, the subjective scores can be replaced by the synthesized quality-aware values using some statistical mathematical methods or well-known FR IQA methods in the pre-training stage. For example, Ma et al. (2017) established a large-scale database named the Waterloo Exploration database, which consists of 94,880 distorted images created from 4744 pristine images, as depicted in Fig. 7.12. Three alternative test criteria were established to evaluate the performance of 2D IQA algorithms (i.e., D-test, L-test and P-test) instead of subjective scores. Motivated by this, Liu et al. (2017) designed a new strategy that the ranked image pairs were generated synthetically on Waterloo Exploration database according to the rank rule that higher distortion levels indicate lower image quality. Using the pairs of the ranked images, a Siamese network (Chopra et al. 2005) was first pre-trained to learn distortion levels of images, and then a branch of the proposed Siamese network was fine-tuned to predict the subjective scores, which aimed to transfer the targets from image distortion levels to

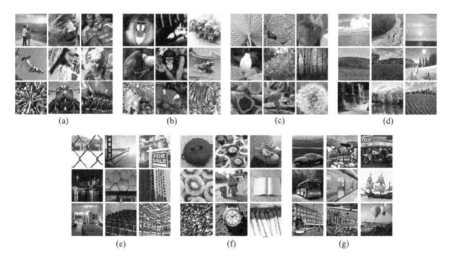

Fig. 7.12 Sample source images in Waterloo exploration database, including seven categories: **a** human, **b** animal, **c** plant, **d** landscape, **e** cityscape, **f** still-life and **g** transportation

quality scores (Liu et al. 2017). In Sect. 7.3, it is considered a pretty good idea that the generated local average quality map is recognized as the temporary ground-truth annotations when adopting the patch-wise strategy. It seems feasible to create a larger image dataset with quality-aware labels for 2D IQA tasks. Following the construction procedures of Waterloo Exploration database, Dendi et al. (2019) constructed an image dataset generated from 120 natural images and introduced four common distortion types including JPEG compression, JP2K compression, White Noise (WN) and Gaussian Blur (GB) with five distortion levels. These total 2400 distortion images are used to train a convolutional autoencoder (CAE) model with their corresponding SSIM distortion maps as ground-truth labels, achieving promising results. Nevertheless, although it seems to be a feasible approach, there is no one algorithm applying the insights for SIQA tasks to our knowledge.

7.6 Other Necessary Knowledge in CNN-Based SIQA Tasks

In addition to various CNN architectures, there are some necessary knowledge points need to be understood before training deep models, which can be divided into five aspects: image preprocessing, activation function, loss function, regularization and optimization.

7.6.1 Image Preprocessing

Image preprocessing plays an essential role in deep learning before training a CNN model. This is because that a more canonical form of both training and testing sets by reducing the amount of variation of CNN inputs can increase both the generalization and robustness of deep models, and even converge only by using a small canonical dataset. Image preprocessing has been applied into many computer vision tasks, which can be recognized as a necessary measure especially for the small training dataset. When training a deep model on a small dataset, necessary preprocessing methods should be adopted to remove some kinds of variability in the image inputs. For IQA tasks, Mittal et al. (2012) have proved that the statistical properties of local normalized contents are closely sensitive to the corruption of images. By applying the local normalization, the objective function can easily reach the global minima of the IQA tasks. The image preprocessing procedure can be obtained by the following function:

$$I'(i, j) = \frac{I(i, j) - \mu(i, j)}{\sigma(i, j) + C}$$

$$\mu(i,j) = \sum_{k,l} \omega_{k,l} I_{k,l}(i,j)$$

$$\sigma(i,j) = \sqrt{\sum_{k,l} \omega_{k,l}(I_{k,l}(i,j) - \mu(i,j))^2} \tag{7.10}$$

where $I(i,j)$ is a given intensity image input, ω demonstrates a two-dimensional circularly symmetric gaussian weighting function. u and σ represent the mean and variance of the inputs, respectively. C is a small positive constant for avoiding the denominator being zero.

However, for the large datasets and complex model training, this kind of image preprocessing is considered unnecessary. For example, Krizhevsky et al. (2012) only adopt one preprocessing step (i.e., subtracting the mean across training data of each pixel) to train AlexNet on ImageNet for image classification. Therefore, which methods of image preprocessing depends on the tasks and the designed models.

7.6.2 Activation Function

When designing a CNN model into IQA fields, a proper activation function can significantly improve the prediction performance of image quality. In the early studies, the optimization of neural networks mainly adopted sigmoid function or tanh function, whose output was bounded and could easily serve as the input of the next layer. However, gradient vanishing or gradient exploding could occur in the procedure of back propagation using sigmoid or tanh function because the accumulation of the gradients through these activation functions easily tends to zero or very large.

To alleviate the problem of gradient vanishing, rectified linear unit (ReLU) is first proposed in Nair and Hinton (2010), which is one of the most notable non-saturated activation functions. The activation function can be expressed by:

$$y_i = \max(0, z_i) \tag{7.11}$$

where z_i is the input of i-th CNN channel. The simple *max* operation of ReLU allows it to compute much faster than the introduced sigmoid or tanh activation function. At the same time, it is also likely to obtain sparse representations in the hidden units. However, there exist some disadvantages of ReLU, that is the output is non-zero-centered. When the input unit is not active, ReLU will force the output to be zero, leading to the gradient of this part is always equal to zero.

Another common activation function, called leaky ReLU (LReLU) (Maas et al. 2013), is introduced to alleviate this problem, which is defined as:

$$y_i = \begin{cases} z_i, & z_i \geq 0 \\ \alpha z_i, & z_i < 0 \end{cases} \tag{7.12}$$

where α is a hyper parameter for controlling the output ratio when the unit is not active. LReLU allows a small, non-zero gradient when the input is less than 0 by compressing the negative part instead of forcing it to zero.

The improvement of LReLU is parametric ReLU (PReLU) (He et al. 2015), which adaptively learns the parameters of the rectifiers instead of using a fixed hyper parameter. The parameters of the rectifiers can be trained simultaneously with other parameters by backpropagation to improve accuracy.

In addition, there are some newly common activation functions suitable for different applications, such as exponential linear unit (ELU), maxout and so on. Limited by the length of the book, we will not elaborate here in detail. The readers interested of this can read some relevant materials (Clevert et al. 2015; Goodfellow et al. 2013).

7.6.3 Loss Function

In CNN-based IQA schemes, it is important to choose an appropriate loss function for optimizing networks. In general, for regression tasks, two common loss functions, namely mean absolute error (MAE) and mean square error (MSE), are usually utilized as minimization criterion, where the curve graph is shown in Fig. 7.13. MSE is the mean square of the distance between the predicted value of the deep model $f(x_i)$ and the ground-truth value y of samples x_i as follows:

$$MSE = \frac{1}{m} \sum_{i=1}^{m} (y_i - f(x_i))^2 \qquad (7.13)$$

where m represents the total number of training data samples.

In the similar way, the definition of MAE can be expressed by the following function:

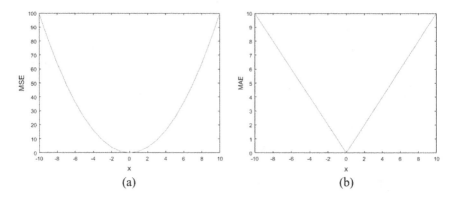

Fig. 7.13 The curve of MSE and MAE loss functions: **a** MSE, **b** MAE

7.6 Other Necessary Knowledge in CNN-Based SIQA Tasks

$$MAE = \frac{1}{m} \sum_{i=1}^{m} |y_i - f(x_i)| \qquad (7.14)$$

where $f(x_i)$ denotes the predicted score from the designed deep model. y_i is the subjective quality score of the image x_i. m represents the total number of training data samples.

The advantages of MSE are smooth, continuous, differentiable and easy to converge, which is a relatively common loss function for regression tasks. Moreover, even if the learning rate is fixed, the function can quickly obtain the minimum value when $y_i - f(x_i)$ less than 1. However, with the error increasing between predicted value and ground-truth value, MSE will lead to a loss value. In another words, MSE will give a larger penalty for the case with a larger error in the training procedure. If there exist outliers in the training data, MSE will assign higher penalty weights to these outliers, resulting in the normal training data being assigned smaller weights, which will eventually decrease the overall prediction performance of the deep model. Compared with MSE, MAE has the same gradient value in most cases, which means that MAE is less sensitive to outliers. Nevertheless, for the case of $y - f(x) = 0$, derivatives don't exist in MAE, and meanwhile a larger penalty will be given even if the small loss values. This is not conducive to the convergence of loss functions and the learning of deep models.

To integrate the advantages of MSE and MAE, Huber loss is designed as follows:

$$L_\delta(y_i, f(x_i)) = \begin{cases} \frac{1}{2}(y_i - f(x_i))^2, & |y_i - f(x_i)| \leq \delta \\ \delta |y_i - f(x_i)| - \frac{1}{2}\delta^2, & |y_i - f(x_i)| > \delta \end{cases} \qquad (7.15)$$

where the hyper-parameter δ determines the emphasis of Huber loss on MSE and MAE. When the difference between y_i and $f(x_i)$ is small ($<\delta$), Huber loss can be recognized as MSE. With the increasing of $y_i - f(x_i)$ ($>\delta$), Huber loss is equal to MAE.

In addition, for classification tasks, many loss functions have been designed and proved to achieve promising results, such as softmax loss, hinge loss and contrastive loss. Taking softmax loss as an example, the models proposed in Ding et al. (2018), Xu et al. (2019) were trained using softmax loss as the loss function. More information can be found in the relevant paper and we will not describe here in detail.

7.6.4 Regularization

Overfitting is usually encountered in the training procedure of complex CNN networks, especially for IQA tasks, a regression task with small samples. Experiments have proved that adopting some technologies of regularization can effectively reduce the degree of overfitting to some extent. Here we will introduce three effective regularization techniques already applied into CNN-based IQA tasks.

1. *Dropout*: Dropout is first introduced by Hinton et al. (2012) used for addressing the problem of overfitting caused by the complex CNN architecture. The key idea is to randomly drop units from FC layers during training, which can prevent the networks from becoming too dependent on any units of FC layers. During training, dropout samples from an exponential number of different "thinned" networks. During the procedure of testing, the prediction results of the networks are accurate using a single unthinned network even if lack of certain information. In recent years, several methods have been proposed to improve the performance of dropout, which can be found in relevant papers (Kang et al. 2014; Kim and Lee 2017).
2. *L1/L2 Regularization*: Regularization has been applied for decades, in which L1/L2 norms are two of the simplest and most popular regularization. Let $\Omega(\theta)$ represents the norm penalty term of L1 or L2 norm and L is the standard objective function. The target function can be expressed by:

$$\theta^* = \arg\min_{\theta} \sum_i L(y_i, f(x_i; \theta)) + \lambda \Omega(\theta) \quad (7.16)$$

where λ is the hyper parameter, and larger values of λ represent larger regularization.

L1 norm is a common method to penalize the size of the model parameters, whose corresponding norm penalty term is defined as:

$$\Omega(\theta) = \|\omega\| = \sum_i |\omega_i| \quad (7.17)$$

Different from L1 norm, the norm penalty term of L2 norm can be expressed as the square of the model parameters ω. By optimizing the designed norm penalty term, L1 norm is more likely to obtain sparse model parameters, while L2 norm drives the model parameters tending to zero.
3. *Data Augmentation*: Data Augmentation is one of excellent ways to increase the representation robustness and reduce the generalization error of deep models. Common methods consist of flipping, cropping, rotations, reflections, and so on, where most of them have been proved to be effective in most computer vision fields by creating more data samples before training.
4. *Batch Normalization*: Batch normalization is proposed by Ioffe and Szegedy (2015), which is designed for accelerating the training process of deep neural networks. It accomplishes the so-called covariate shift problem by a normalization step that fixes the means and variances of layer inputs where the estimations of mean and variance are computed after each mini-batch rather than the entire training set. Moreover, the small disturbance of inputs distribution also can be considered as an effective method in preventing overfitting due to its regularization effects in data distributions, like data augmentation.

7.6.5 Optimization

The last aspect we would like to discuss is optimization. The neural network model is optimized by gradient descent algorithm that follows the gradient of an entire training dataset downhill. The optimization can be realized by several common and effective algorithms, including stochastic gradient descent (SGD), momentum (Qian 1999), nesterov momentum (Sutskever et al. 2013), AdaGrad (Duchi et al. 2011), RMSProp (Hinton et al. 2012), adaptive moments (Adam) (Kingma and Ba 2015) and so on, where SGD and Adam will be introduced in this subsection in detail.

SGD is probably the most used and fundamental algorithm to optimize CNN models for better convergence. Comparing with gradient descent, the insight of SGD is that the gradient is estimated approximately using the average of a small set of samples, called minibatch. The optimization procedure of SGD can be recognized as the following function:

$$\theta^* = \theta - \eta \frac{1}{m} \nabla_\theta \sum_{i=1}^{m} L(y_i, f(x_i; \theta)) \qquad (7.18)$$

where m represents the number of minibatch generally set to be a relatively small number of examples.

In deep learning research, the set of learning rate is indeed one of the most difficult problems. To solve that problem, the insight of adjusting the learning rate adaptively during training is considered feasible. More recently, a number of adaptive learning rate optimization algorithms have been introduced in the optimization, such as AdaGrad, RMSProp and Adam, where Adam (Kingma and Ba 2015) is recognized as a master in optimization algorithms. Adam directly integrates momentum and RMSProp as the estimates of the first-order moments and second-order moments to accelerate the training procedure and to reach the optimal hyperplane.

For the question of choosing a proper optimization algorithm for a specific CNN model in a specific task, there is currently no consensus on this point. Schaul et al. (2014) have conducted the comparison experiments of a large number of optimization algorithms on a series of deep learning tasks. The results of experiments demonstrated that there is no single best algorithm performing well on all learning tasks, although the family of algorithms with adaptive learning rates could perform more fairly robustly compared with those with fixed learning rates. However, the other experiment concluded the opposite conclusion that SGD and SGD with momentum outperform adaptive methods on the test dataset (Wilson et al. 2017). Thus, which optimization algorithm to use seems to depend largely on ourself.

For more information about optimization, please see Ruder (2016).

7.6.6 Summary

In the end of this section, for a more illustrative understanding, the key information of the introduced CNN-based SIQA models is summarized in Table 7.2. Several important observations are made: (1) All the proposed CNN-based SIQA models were equipped ReLU activation function in convolutional layers, that's because there always remain positive values during training due to its non-negative ground-truth labels. In addition, Krizhevsky et al. (2012) demonstrated that ReLU enables to train the deep CNN models faster than that using tanh or sigmoid activation function. (2) For different tasks, there are different loss functions to be chosen. For example, Ding et al. (2018) and Xu et al. (2019) selected the softmax loss as loss function because they recognized the tasks as quality classification. On the contrary, MAE or MSE would be applied into the quality regression tasks to expected to obtain an accuracy predicted value. (3) Limited of lack of adequate training data, regularization techniques were generally adopted in CNN-based SIQA models for preventing overfitting, such as dropout, L2 norm and batch normalization. (4) As for the optimization approaches, there is no clear and unified scheme, where Adam can complete the optimization target easily and quickly, while SGD with proper hyper parameters is likely to achieve satisfactory optimization results.

7.7 Summary and Future Work

Obviously, significant progresses have been made in both CNN-based 2D and 3D IQA tasks in the last three or four years. To summarize them, as well as to offer some prospects for the future, some related discussions and further analysis will be given as follows.

When employing CNNs into IQA fields, a critical obstacle must be faced and solved, which is lack of training data. As already discussed in previous sections, constructing large-scale training datasets for IQA is a much more difficult problem than other image processing applications, like image recognition and classification. Either creating distorted images or assigning corresponding subjective scores require time-consuming and expensive subjective studies. Especially for obtaining ground-truth quality labels, it must be conducted under controlled laboratory conditions, which makes this approach less possible. A common strategy is data augmentation technique suitable for many image processing tasks, which can multiply the number of training data via cropping, rotations, and so on. Unfortunately, due to the strong relationship between image information and the corresponding subjective score, these data augmentation techniques cannot be applied into IQA tasks except for horizontal flip. In addition, unlike other image processing tasks, reliable subjective label for each image is not easy to obtain because the collection of subjective MOS values is a complex and time-consuming procedure, meaning that creating more distorted image datasets evaluated by subjective quality assessment is not possible.

7.7 Summary and Future Work

Table 7.2 Key information about the databases

Models	Type	Structure	Preprocessing	Activation function	Loss function	Regularization	Optimization
Ding et al. (2018)	Transfer learning	AlexNet	N/A*	ReLU	Softmax	N/A	SGD
Xu et al. (2019)		AlexNet, GoogLeNet	N/A	ReLU	Softmax	N/A	SGD with momentum
Zhang et al. (2016)	Patch-wise	2 Conv layers, 2 FC layers	Local normalization	NAN	MSE	Dropout	SGD
Oh et al. (2017)		2 Conv layers, 4 FC layers	Local normalization	ReLU	MSE	Batch normalization	Adam
Sun et al. (2020)		VGG16	Local normalization	ReLU	MAE	Dropout, L2 norm	Step 1: Adam, Step 2: SGD with momentum
Zhou et al. (2019)		5 Conv layers, 2 FC layers	Local normalization	ReLU	MSE	Dropout	SGD with momentum

*N/A means the unavailable information which is not introduced in the original paper

In another strategy, CNN models can be trained by splitting images into several fixed-size patches, namely the patch-wise strategy. However, the strategy also encounters the same problem: subjective score is not available for each patch. There are two methods to obtain the ground-truth label for each patch, respectively assigning the subjective quality score to each patch and generating the local average SSIM value as the training label of each patch. Experiments have proved that the latter is more likely to achieve good prediction performance. That's because the distortions of images are not homogeneous, and not highly consistent with human perceptions, leading to the fact that the overall subjective score of the image differs with the one of each image patch generated from the entire image. The last but not least, although creating more training data faces the tricky problem of lacking subjective labels, we can adopt classical FR IQA methods to generate quality maps as the proxy ground-truth maps, where the generated labels are ensured to sensitive to image quality.

Different from 2D natural images, stereopairs contain richer 3D visual information, such as binocular vision, disparity information. How to design a CNN model to learn binocular features from original left and right views is also a very difficult problem except for insufficient training data. In general, a straightforward way is to feed the synthesized cyclopean images into the CNN models designed for 2D IQA tasks and obtain the output as predicted quality score. Another strategy for SIQA tasks is to design multi-column CNN model, where each CNN branch is designed to learn visual features for monocular or binocular view, respectively. Then all visual features learned from every CNN branch are concatenated and aggregated into the final objective rating. However, there is no one method including these introduced before can be recognized as a perfect solution. That's because it is hard to model the complex binocular visual properties from simple and complex cells in human brain.

To summary, due to the late start of CNN-based SIQA research, and limited by the problem of inadequate data sets and complex visual system, there are not many studies on CNN-based SIQA tasks. However, inspired by the progress of other stereo image processing tasks (e.g., object detection, stereo matching), deep learning techniques are likely to offer satisfactory performance and breakthroughs for objective SIQA research in the near future.

References

Bengio Y (2011) Deep learning of representations for unsupervised and transfer learning. In: ICML unsupervised and transfer learning challenge workshop, pp 17–36

Bianco S, Celona L, Napoletano P, Schettini R (2018) On the use of deep learning for blind image quality assessment. Sig Image Video Process 12(2):355–362

Chan T-H, Jia K, Gao S, Lu J, Zeng Z et al (2015) PCANet: a simple deep learning baseline for image classification. IEEE Trans Image Process 24(12):5017–5032

Chen MJ, Su CC, Kwon DK, Cormack LK, Bovik AC (2013) Full-reference quality assessment of stereopairs accounting for rivalry. Sig Process Image Commun 28(9):1143–1155

References

Cheng Z, Takeuchi M, Katto J (2017) A pre-saliency map based blind image quality assessment via convolutional neural networks. In: Proceedings of 2017 IEEE international symposium on multimedia (ISM), Taichung, pp 77–82

Chopra S, Hadsell R, LeCun Y (2005) Learning a similarity metric discriminatively, with application to face verification. In: Proceedings of the IEEE conference on computer vision and pattern recognition, pp 539–546

Clevert DA, Unterthiner T, Hochreiter S (2015) Fast and accurate deep network learning by exponential linear units (elus). In: Proceedings of the international conference on learning representations, San Diego, CA, USA

Dendi SVR, Dev C, Kothari N, Channappayya SS (2019) Generating image distortion maps using convolutional autoencoders with application to no reference image quality assessment. IEEE Sig Process Lett 26(1):89–93

Ding Y, Deng R, Xie X, Xu X, Chen X, Krylov AS (2018) No-reference stereoscopic image quality assessment using convolutional neural network for adaptive feature extraction. IEEE Access 6:37595–37603

Dou P, Shah SK, Kakadiaris IA (2017) End-to-end 3D face reconstruction with deep neural networks. In: Proceedings of the IEEE conference on computer vision and pattern recognition, Hawaii, USA, pp 1503–1512

Duchi J, Hazan E, Singer Y (2011) Adaptive subgradient methods for online learning and stochastic optimization. J Mach Learn Res 12:2121–2159

Gao F, Wang Y, Li P, Tan M, Yu J, Zhu Y (2017) DeepSim: deep similarity for image quality assessment. Neurocomputing 104–114

Goodfellow IJ, Warde-Farley D, Mirza M, Courville A, Bengio Y (2013) Maxout networks. In: Proceedings of the international conference on machine learning, Atlanta, Georgia, pp 1319–1327

He K, Zhang X, Ren S, Sun J (2015) Delving deep into rectifiers: surpassing human-level performance on imagenet classification. In: Proceedings of the international conference on computer vision, Santiago, Chile, pp 1026–1034

He K, Zhang X, Ren S, Sun J (2016) Deep residual learning for image recognition. In: Proceedings of the IEEE conference on computer vision and pattern recognition, Las Vegas, Nevada, USA, pp 770–778

Hinton GE, Srivastava N, Krizhevsky A, Sutskever I, Salakhutdinov RR (2012) Improving neural networks by preventing coadaptation of feature detectors. arXiv:1207.0580

Hou W, Gao X (2015) Saliency-guided deep framework for image quality assessment. IEEE Multimedia 22(2):46–55

Hou W, Gao X, Tao D, Li X (2015) Blind image quality assessment via deep learning. IEEE Trans Neural Netw Learn Syst 26(6):1275–1286

Ioffe S, Szegedy C (2015) Batch normalization: accelerating deep network training by reducing internal covariate shift. J Mach Learn Res 448–456

Jia Y, Shelhamer E, Donahue J, Karayev S, Long J et al (2014) Caffe: convolutional architecture for fast feature embedding. In: Proceedings of 22nd ACM international conference on multimedia, Orlando, Florida, USA, pp 675–678

Jiang Q, Duan F, Shao F (2014) 3D visual attention for stereoscopic image quality assessment. J Softw 9(7):1841–1847

Kang L, Ye P, Li Y, Doermann D (2014) Convolutional neural networks for no-reference image quality assessment. In: Proceedings of the IEEE conference on computer vision and pattern recognition, Columbus, America, pp 1733–1740

Kim J, Lee S (2017) Fully deep blind image quality predictor. IEEE J Sel Top Sig Process 11(1):206–220

Kingma DP, Ba J (2015) Adam: a method for stochastic optimization. In: Proceedings of international conference on learning representations, San Diego, CA, USA, pp 1–13

Krizhevsky A, Sutskever I, Hinton GE (2012) ImageNet classification with deep convolutional neural networks. In: Proceedings of advances in neural information processing systems, pp 1097–1105

LeCun Y, Bottou L, Bengio Y, Haffner P (1998) Gradient-based learning applied to document recognition. Proc IEEE 86(11):2278–2324

Li J, Duan L-Y, Chen X, Huang T, Tian Y (2015) Finding the secret of image saliency in the frequency domain. IEEE Trans Pattern Anal Mach Intell 37(12):2428–2440

Li Y, Po LM, Feng L, Yuan F (2016) No-reference image quality assessment with deep convolutional neural networks. In: IEEE International conference on digital signal processing, Beijing, China, pp 685–689

Liu X, van de Weijer J, Bagdanov AD (2017) RankIQA: learning from rankings for no-reference image quality assessment. In: Proceedings of the international conference on computer vision, Venice, Italy, pp 1040–1049

Lv Y, Yu M, Jiang G, Peng Z, Chen F (2016) No-reference stereoscopic image quality assessment using binocular self-similarity and deep neural network. Sig Process Image Commun 47:346–357

Ma K, Duanmu Z, Wu Q, Wang Z, Yong H et al (2017) Waterloo exploration database: new challenges for image quality assessment models. IEEE Trans Image Process 26(2):1004–1016

Maas AL, Hannun AY, Ng AY (2013) Rectifier nonlinearities improve neural network acoustic models. In: Proceedings of international conference on machine learning, Atlanta, USA

Mittal A, Moorthy A, Bovik A (2012) No-reference image quality assessment in the spatial domain. IEEE Trans Image Process 21(12):4695–4708

Moorthy AK, Su CC, Mittal A, Bovik AC (2013) Subjective evaluation of stereoscopic image quality. Sig Process Image Commun 28(8):870–883

Nair V, Hinton GE (2010) Rectified linear units improve restricted boltzmann machines. In: Proceedings of the 27th international conference on machine learning, Haifa, Israel, pp 807–814

Oh H, Ahn S, Kim J, Lee S (2017) Blind deep S3D image quality evaluation via local to global feature aggregation. IEEE Trans Image Process 26(10):4923–4935

Pan SJ, Yang Q (2009) A survey on transfer learning. IEEE Trans Knowl Data Eng 22(10):1345–1359

Ponomarenko N, Jin L, Leremeiev O, Lukin V, Egiazarian K et al (2015) Image database TID2013: peculiarities, results and perspectives. Sig Process Image Commun 30:55–77

Qian N (1999) On the momentum term in gradient descent learning algorithms. Neural Netw Official J Int Neural Netw Soc 12(1):145–151

Ruder S (2016) An overview of gradient descent optimization algorithms. arXiv: Learning

Schaul T, Antonoglou I, Silver D (2014) Unit tests for stochastic optimization. In: Proceedings of international conference on learning representations, Banff, Canada

Sheikh HR, Sabir MF, Bovik AC (2006) A statistical evaluation of recent full reference image quality assessment algorithms. IEEE Trans Image Process 15(11):3440–3451

Shelhamer E, Jonathan L, Darrell T (2017) Fully convolutional networks for semantic segmentation. IEEE Trans Pattern Anal Mach Intell 39(4):640–651

Simonyan K, Zisserman A (2015) Very deep convolutional networks for large-scale image recognition. In: Proceedings of international conference on learning representations, San Diego, CA, USA

Sun G, Ding Y, Deng R, Zhao Y, Chen X et al (2018) Stereoscopic image quality assessment by considering binocular visual mechanisms. IEEE Access 6:511337–511347

Sun G, Shi B, Chen X, Krylov AS, Ding Y (2020) Learning local quality-aware structures of salient regions for stereoscopic images via deep neural networks. IEEE Trans Multimedia 1

Sutskever I, Martens J, Dahl G, Hinton G (2013) On the importance of initialization and momentum in deep learning. In: Proceedings of the 30th international conference on machine learning, Atlanta, Georgia, USA, pp 1139–1147

Szegedy C, Liu W, Jia Y, Sermanet P, Reed S et al (2015) Going deeper with convolutions. In: Computer vision and pattern recognition, Boston, America, pp 1–9

References

Wilson AC, Roelofs R, Stern M, Srebro N, Recht B (2017) The marginal value of adaptive gradient methods in machine learning. In: Neural information processing systems, Long Beach, California, USA, pp 4148–4158

Xu X, Shi B, Gu Z, Deng R, Chen X et al (2019) 3D no-reference image quality assessment via transfer learning and saliency-guided feature consolidation. IEEE Access 7:85286–85297

Yang J, Zhao Y, Zhu Y, Xu H, Lu W et al (2019) Blind assessment for stereo images considering binocular characteristics and deep perception map based on deep belief network. Inf Sci 474:1–17

Ye P, Kumar J, Kang L, Doermann D (2012) Unsupervised feature learning framework for no-reference image quality assessment. In: Proceedings of IEEE conference on computer vision and pattern recognition, Providence, Rhode Island, pp 1098–1105

Zhang W, Zhang Y, Ma L, Guan J, Gong S (2015) Multimodal learning for facial expression recognition. Pattern Recogn 48(10):3191–3202

Zhang W, Qu C, Ma L, Guan J, Huang R (2016) Learning structure of stereoscopic image for no-reference quality assessment with convolutional neural network. Pattern Recogn 59:176–187

Zhou W, Chen Z, Li W (2019) Dual-stream interactive networks for no-reference stereoscopic image quality assessment. IEEE Trans Image Process 28(8):3946–3958

Chapter 8
Challenging Issues and Future Work

Abstract The complete development history of Stereoscopic Image Quality Assessment (SIQA) has been overviewed in the previous chapters. Even if the state-of-art SIQA methods have achieved competitive results, there still have some challenges and obstacles that need to be discussed and concluded in the end of this book, including subjective studying, in-depth research of Human Visual System (HVS) and the bottleneck of insufficient training data. Finally, the practical applications of SIQA are discussed.

Keywords Stereoscopic image quality assessment · Subjective studying · Human visual system

So far, we have overviewed advanced progresses and main challenges of the fields of stereoscopic image quality assessment (SIQA) in above chapters. Among these state-of-the-art SIQA algorithms, there can be classified into three categories: the first category of SIQA algorithms attempted to recognize the IQA on stereoscopic images as a 2D IQA task and adopted classical 2D IQA methods into each individual view of a stereopair. Then the final objective quality score can be obtained through the average of monocular scores of both the two views. This type of SIQA algorithms is rather simple, but not very effective because the intersection relationship between the left and right views is not be taken into consideration. The second type of SIQA algorithms considers 3D quality factors (e.g., depth perception) in the basis of the first type of algorithms to improve the prediction performance. Lastly, by exploring the interaction of two views in the human visual system (HVS), the third type of studies begins to focus on the generation of a 'cyclopean' image from two eyes through binocular behaviors of simple and complex cells in the human brain. Recently, with the widely used of machine learning and deep learning technologies in image processing, SIQA based on machine learning or deep learning becomes the other hot topic of the research, driving the diversified development of the stereoscopic image quality assessment.

In addition, there still have several obstacles and challenges that need to be discussed and concluded in the end of this book. These challenges could appear in all aspects of the field of SIQA, which are instructive for the SIQA research in the near future. In the following sections, we would like to introduce these obstacles and challenges in detail and consider how they can offer correct research directions for our future work.

The first aspect we would like to introduce is subjective studying. As is known to all, when it comes to subjective studying, it is a very effective and accurate way to predict the quality of a distorted stereoscopic image. However, there are exist some fatal drawbacks in subjective studying: time-consuming, labor-intensive and not very real-time, which can't be widely used in image processing systems. Fortunately, some objective SIQA algorithms with good prediction performance can be used instead of subjective studying. Yet subjective studying is still unavoidable because subjective studying is fundamental to explore the task of objective SIQA, which can be discussed from two aspects. To begin with, the appearance of subjective databases using the technologies of subjective studying extremely drives the formulation and development of objective SIQA. Given the distorted images and its corresponding subjective quality scores, researchers can easily design objective SIQA algorithms to fit the relationship between them, which is the original idea of objective SIQA. On the other hand, with the development of SIQA research based on deep learning in recent years, the inadequate of SIQA databases has gradually become the biggest bottleneck of the research. That's because that SIQA based on deep learning mainly utilizes adaptive learning in massive amounts of labeled data to achieve the purpose of accurate prediction of distorted degree for images. Under this situation, more SIQA database resources need to be developed and established. However, we have known that subjective studying is a time-consuming and laborious processing, making the establishment of subjective databases more difficult. Therefore, how to simplify the process of subjective studying and create more subjective databases with less time and money costs is an issue that needs urgent attention.

As discussed above, limited to the disadvantages of subjective studying, researchers had attempted to begin to the study of objective SIQA and made considerable progress in the field of objective SIQA. According to the contents discussed in Chap. 6, a fact has been declared that the feature extraction of quality perception has been one of the most traditional topics in the history of research from the time of the birth of the IQA project to the research of SIQA. It is worth noting that the selection of quality-ware features is not random, but rather based on careful consideration of the established reality, that is in line with the perceived characteristics of HVS. According to previous works, various types of quality-aware features, such as brightness, contrast, edges, textures, etc., have been shown as effective tools to represent image quality, thereby achieving performance improvements of IQA algorithms. Motivated by this, more and more IQA studies focus on visual feature extractions exhibited by HVS in the human brain. On the one hand, IQA researchers have explored inherent characteristics of HVS, in which several low-level visual features (e.g., edges, spatio-temporal frequencies) and high-level visual features (e.g., textures, shapes and color information) have been successively established and applied to IQA research, effectively enhancing the robustness of the IQA algorithm. On the other hand, more complex psychological visual characteristics such as visual attention mechanism and multi-scale energy response have also been shown to help improve prediction performance of the SIQA algorithm. In addition, stereo vision also becomes the other main obstacle when it comes to SIQA research. Therefore, for improving the performance of SIQA research, more in-depth research and understanding of HVS including psychological and physiological research is necessary.

Third, the popularity of machine learning technology has greatly affected SIQA research. The main idea of machine learning is to optimize the parameters of the prediction model a through the training of a large amount of priori knowledge. In IQA research, the theory of machine learning was first used in the mapping stage, that is, to establish a mapping relationship between the extracted visual feature vector (usually several values) and the target value (which is usually one value) to obtain a regression model. The application of naive regression models such as linear regression and exponential regression can be found in the early research stages of IQA. In recent years, with the development of machine learning technology, complex regression models such as support vector machine (SVM) and neural network (NN) have been widely used due to the significant improvement in prediction performance. However, with the regression model becomes more and more complex, the training procedure of regression model requires more and more training samples. Therefore, the trade-off between model accuracy and complexity has attracted the attention and thought of researchers. On the other hand, another important application of machine learning is to replace the role of feature extraction. Although currently the training data is far from sufficient to "learn" how to extract visual features automatically, convolutional neural network (CNN) applied in object recognition provides a good example of this possibility. Therefore, SIQA based on machine learning is still a subject worth studying.

Since it was invented in 1998, CNN has been attracting a lot of interest of researchers working in computer vision and artificial intelligence. Compared to machine learning discussed in the previous paragraph, CNN can achieve more accurate prediction performance, especially for the field of image processing. Of course, the performance improvement of CNN comes at the cost of model complexity, that is, more training data is needed to obtain better prediction results. Nevertheless, some researchers have applied CNN to the research of SIQA. When employing a CNN architecture directly for SIQA tasks, the critical obstacle of insufficient data needs to be taken seriously. In the existing CNN-based SIQA methods, transfer learning and data augmentation methods are generally used to alleviate the problem of inadequate training data. Transfer learning can achieve the goal of quality prediction of stereo images by pre-training the CNN model on a large image database such as the ImageNet database and then fine-tuning the last few layers of the CNN model on available SIQA databases. Data augmentation mainly adopts the patch-wise strategy to expand the training set for the purpose of effectively and sufficiently training the CNN model, from which the stereoscopic image will be divided into several non-overlapping image patches. However, both of them have their shortcomings, in which the former cannot guarantee whether the features learned from the pre-training process are sensitive to the degree of corruption or not, and the latter sacrifices the global feature information of stereoscopic images. Recently, there breeds another idea to avoid the problem of insufficient data by directly creating more training data. However, because the collection of human subjective quality scores is a complex, time-consuming procedure, reliable subjective labels of distorted stereopairs are not easily obtained, which has not yet formed an effective SIQA algorithm based on the idea of extended SIQA databases. However, it is undeniable that the application of CNN-based SIQA research can extremely improve the prediction performance. Therefore, when the bottleneck of insufficient training data is broken in the near future, the research on CNN-based SIQA will become a hot spot.

Although the objects we have been discussing throughout in this book are stereoscopic images, thanks to the development and application of virtual reality technology, and omnidirectional images will gradually become the focus of IQA research. Comparing with stereoscopic images, omnidirectional image can provide the viewer with a larger view, including bipolar views and equatorial view. When bipolar regions are distorted, the distortion of omnidirectional image will be magnified, affecting greatly the viewers' experience. However, to the best of our knowledge, the research of omnidirectional image quality assessment had been a largely under explored domain. In addition, the rapid development of 3D video applied in 3D cinemas and home theaters also drive the study of 3D video contents. Similar to 3D images, it will also lead to the degradation of 3D video during the processing, compression and transmission of 3D video. Therefore, the research for 3D video quality assessment also is one of the main directions of future objective SIQA study.

Last but not least, all theoretical research must have its practical application to reflect the value of research. The same conclusion is suitable for the SIQA research. We have already known that IQA has been implemented in many application-specific fields, such as medical image screening for medical imaging and real-time monitoring of image processing systems. SIQA can also have similar situations in many fields, such as image quality monitoring of 3D image transmission systems and image optimization of camera systems of mobile phones. The development of a dedicated method is a very good sign of SIQA research, as its practical usages has been confirmed. The practical application of SIQA can involve the detailed processing scheme implemented by SIQA algorithm. More importantly, in application-specific SIQA, computational complexity becomes a very essential indicator for SIQA research, which is usually ignored in SIQA research. Of course, in the long run, SIQA in each specific application scenario would possibly be enlightening for each other, and make SIQA research more advanced.

To sum up, SIQA has been evolved a lot since the first ten years of twenty-first century as a new research object. In the basis of advancing progress in IQA research, researchers can easily set up a series of standard experimental flows of SIQA studies, including subjective image databases and evaluation standards. In the establishment of SIQA algorithms, quality-aware visual features are designed, properties of HVS are made use of, the models of binocular combination are constructed, the technologies of deep learning are utilized, the performance of objective methods is improved significantly. In practical application, SIQA is generally recognized as an effective monitoring tool in the modern image and video processing systems, and can be equipped for practical usages in many specified image processing areas. However, as far as we are concerned, there are still room for improvement. The "gap" between the current status of SIQA and the subjective judgements still exists, and how to narrow this gap is the focus of SIQA research in the foreseeable future.

CPSIA information can be obtained
at www.ICGtesting.com
Printed in the USA
BVHW010216051120
592594BV00001B/3

9 789811 577635